Archaeological Approaches
to Technology

Archaeological Approaches *to* Technology

Heather Margaret-Louise Miller

Walnut Creek, California

LEFT COAST PRESS, Inc.
1630 North Main Street, #400
Walnut Creek, CA 94596
www.LCoastPress.com

Copyright © 2009 by Left Coast Press, Inc.

First paperback edition 2009.
Hardback edition originally published in 2007 by Academic Press,
an imprint of Elsevier, under ISBN 0-12-496951-8

All illustrations © Heather M.-L. Miller, except the following, approved for use in this volume: Figure 2.3 © Patrick Lubinski; Figures 3.3, 3.7, 4.13 and 4.14 © Lisa Ferin; Figures 3.22, 3.24 and 3.25 © Roger Ivar Lohmann; Extracted translation text in Chapter 3 © Roger Ivar Lohmann.

All rights reserved. No part of this publication may be reproduced, stored in a retrieval system, or transmitted in any form or by any means, electronic, mechanical, photocopying, recording, or otherwise, without the prior permission of the publisher.

ISBN 978-1-59874-473-6 hardcover
ISBN 978-1-59874-474-3 paperback

Library of Congress Cataloging-in-Publication Data

Miller, Heather M. L.
      Archeological approaches to technology/Heather M. L. Miller.
         p. cm.
1. Social archaeology. 2. Technology–Social aspects–History.
3. Prehistoric peoples. 4. Material culture–History.
5. Archaeology–Methodology. 6. Archaeology–Field work.
7. Antiquities–Collection and preservation. I. Title
      CC72.4.M554 2007
      930.1028–dc22
      2006036457

Printed in the United States of America

∞™ The paper used in this publication meets the minimum requirements of American National Standard for Information Sciences—Permanence of Paper for Printed Library Materials, ANSI/NISO Z39.48–1992.

09 10 11 12 13   5 4 3 2 1

CONTENTS

List of Figures — ix
Dedication — xiii
Preface and Acknowledgements — xv

1 Introduction: Archaeological Approaches to Technology — 1
Terminology — 3
Archaeology and Technology Studies — 7
Overview of Volume — 9

2 Methodology: Archaeological Approaches to the Study of Technology — 13
Archaeological Field Techniques: Discovery/Recovery — 16
    Survey — 17
    Excavation — 19
The Examination of Archaeological Remains — 21
    Simple Visual Examination and Measurement — 22
    Complex Examination of Physical Structure and Composition — 25
Ordering and Analyzing Data — 27
    Reconstructing Production Processes; *Chaîne Opératoire* — 29
Analogy and Sociocultural Interpretation — 30
    Experimental Archaeology — 34
    Ethnography, Ethnoarchaeology, and Historical Accounts — 36

3 Extractive-Reductive Crafts — 41
Classification of Crafts — 43
Stone/Lithics — 46
    Collection and Preliminary Processing — 47

Shaping and Finishing Methods	54
Knapping	54
Cutting (Sawing, Drilling, Groove-and-Snapping)	57
Pulverizing (Pecking)	58
Abrading (Grinding, Smoothing, Polishing, Drilling)	59
Production Stages	59
Organization of Production; Consumption	61
Fibers: Cordage, Basketry, Textiles	65
Collection and Preliminary Processing of Fibers	68
Production of Strands and Cordage	72
Fabric Production	75
Ornamentation and Joining	81
Organization of Production and Scheduling Demands	85
Wood, Bone, and Other Sculpted Organics (Antler, Horn, Ivory, Shell)	89
Collection and Preliminary Processing	91
Shaping and Finishing Methods	94
Organization of Production; Use and Reuse of Hard Organic Objects	98

4 Transformative Crafts   101

Fired Clay	103
Collection and Preliminary Processing; Formation of the Clay Body	109
Shaping Methods	113
Drying and Surface Treatments	118
Firing	121
Post-Firing Surface Treatments and Second Firings	128
Vitreous Silicates: Glazes, Faiences and Glass	128
Collection and Preliminary Processing	130
Creating the Vitreous Silicate Mixtures; Fritting; Melting of Glass (Glass Making)	135
Shaping of Faience and Glass Objects	138
Application of Glazes to Faience and Other Materials	141
Firing of Faience and Glazed Objects; Annealing of Glass	143
Post-Firing Surface Treatments	144
Metals: Copper and Iron	144
Collection, Including Mining	147
Processing of Ores and Native Copper; Fuel and Fluxes	150
Smelting	152
Refining and Alloying	156
Shaping and Finishing Methods: Casting and Fabrication (Including Forging)	159
Casting	159
Fabrication	162

5 Thematic Studies in Technology — 167

Technological Systems: Reed Boat Production and Use — 168
    Reconstructing Reed Boats and Exchange Networks in the Arabian Sea — 169
    Reconstructing Reed and Plank Boats and Exchange Networks in Coastal Southern California — 173
Innovation and the Organization of Labor — 180
    The Case of the Grain Harvesting Machine — 181
    Divisions of Labor, Women's Roles, Specialization, and Mass Production of Pottery — 185
Technological Style — 191
    Style and Technological Style — 191
    Technological Traditions: Metal and Bone Working in North America — 195

6 Thematic Studies in Technology (*Continued*) — 203

Value, Status, and Social Relations: The Role of New Artificial Materials in the Indus Valley Tradition — 203
    Uses of Artificial Materials — 204
    Status Differentiation and the Development of Vitreous Materials — 206
    Determining Relative Value — 212
    Social Relations and the Relative Value of Indus Talc-Faience Materials — 217
    Artificial Materials and Cultural Value Systems — 225
Technologies of Religious Ritual in the American Southwest — 226
    Religious Mural Construction, Use, and Discard — 228
    Archaeological Identification of Religious Ritual — 232

7 The Analysis of Multiple Technologies — 237

Cross-Craft Perspectives — 237
Technological Style and Cross-Craft Interactions — 239

Bibliography — 247
Index — 283

LIST OF FIGURES

Additional information about the illustrations is provided in this list, which is not provided in the figure captions.

2.1 Sorted fired clay fragments at the Indus urban site of Harappa. The only fragments indicative of high-temperature firing are in the small pile to the far right. 14
2.2 Surface survey, with flags used to mark objects until they are mapped, recorded, and/or collected. (Heather M.-L. Miller and Sayeed Ahmed Haderi at Harappa, Punjab, Pakistan) 18
2.3 Excavation; shovel-skimming the base of a trench in a plowed field. (Patrick Lubinski, American Midwest) *Photo courtesy of Patrick Lubinski.* 20
2.4 Fired clay object types from Harappa, including debris from high-temperature production, nodules, and lower-fired clay cakes and balls. 24
2.5 (a) Pottery kiln remains from Mohenjo-daro, in relation to (b) number and distribution of drips. (Redrawn after Pracchia 1987.) 25
2.6 Example of a generalized production process diagram for copper and iron (greatly simplified). Explained further in Chapter 4. 29
2.7 Example of an idealized site with hearths between buildings. 31
3.1 Generalized production process diagram for stone (greatly simplified). 48
3.2 Products and some debris from stages of production for agate/carnelian beads produced in Khambhat, India; progression of production stages is from left to right. *Materials courtesy of J. Mark Kenoyer.* 51

3.3	Heat treatment of agate nodules inside pottery vessels, using rice husks as fuel in a reducing atmosphere. Inayat Hussain's workshop, Khambhat, India. *Photo courtesy of Lisa Ferin.*	52
3.4	Experimental heat treatment of chert fragments: (a) placed on sand; (b) covered in more sand; (c) buried under dirt and a fire set.	52
3.5	Illustrations of various percussive techniques for knapping: (a) direct percussion with hard hammer; (b) indirect percussion; (c) block and anvil technique; (d) bipolar technique; (e) inverse indirect percussion; (f) pressure flaking. (Redrawn after Kenoyer n.d. and Odell 2004.)	55
3.6	Creation of building stone blocks by knapping, using direct percussion with iron hammers (Vijayanagara, India).	56
3.7	Creation of agate beads by knapping using inverse indirect percussion: (a) Inayat Hussain, mastercraftsman, and (b) workman in the Keseri Singh workshop (Khambhat, India). *Photos courtesy of Lisa Ferin.*	56
3.8	Flake of obsidian created by bipolar technique; note the two points of percussion. *Obsidian and explanation of technique courtesy of Liz Sobel.*	57
3.9	Stages in production of soapstone (talc/steatite) beads (right to left): pieces cut with a copper saw, drilled, strung, and edges abraded as a unit on a wet grinding stone. (Harappa, Pakistan)	58
3.10	(a) Hand drill, (b) bow drill, and (c) pump drill. (Drawn after photographs in Foreman 1978.)	60
3.11	Small stone adze hafted to wooden handle with plant fibers, used by women for weeding gardens, and unhafted large stone adze, used by men for chopping wood, both from central New Guinea. (Small adze from Yakob, Duranmin, and large adze from Sumwari, East Sepik province, Papua New Guinea.)	61
3.12	Roger Lohmann (left) and Salowa Hetalele (right), Yakob village, Duranmin, Papua New Guinea; July 2005, ten years after the interview presented in text.	62
3.13	Generalized production process diagram for fibers, basketry and textiles (greatly simplified).	69
3.14	Tree fibers of *Gnetum gnemon* (Tok Pisin *tu-lip*) hanging to dry after processing by Mandi Diyos (right), prior to thigh-spinning for string bag (bilum) manufacture. (Ilim Oks, Duranmin, PNG) See MacKenzie 1991: 70–73 for detailed processing information.	71
3.15	Making thread by (a) rolling *tu-lip* fibers on thigh (Sansip at Yakob, Duranmin, PNG); (b) spinning wool with a drop spindle (Afghan woman near Harappa, Pakistan); (c) using a spinning wheel with cotton fiber (Rita Wright, left; near Harappa, Pakistan).	73

List of Figures xi

3.16	Examples of net-making tools and portions of fishing nets. *Materials courtesy of William R. Belcher.*	76
3.17	(a) Coiled basket made of grass (Ravi River, Punjab, Pakistan); (b) Woven rectangular basket made from dyed wood strips (made by Velma Lewis, Ho-Chunk tribe, Wisconsin, USA, 1988); (c) Woven circular basket made from twigs (provenance unknown; European or Euro-American style).	77
3.18	Drawings of different types of looms: (a) sketch of back-strap loom; (b) drawing of warp-weighted loom, with schematic of shed formed by shed rod and heddle; (c) lower portion of a horizontal beam-tension loom, showing details of shed formed by shed rod and heddle operation. (Redrawn after Brown 1987 and Hodges 1989 [1976].)	80
3.19	(a) Examples of tie resist dyeing (Gujarat, India, left) and block printing (Sindh, Pakistan, right); (b) example of carved wooden stamps used to print cloth (stamps from Delhi, India).	84
3.20	Generalized production process diagram for wood and other sculpted organics (greatly simplified).	91
3.21	Living cedar trees from Northwest Coast of North America (British Columbia) with planks removed. (Redrawn after Stewart 1984.)	93
3.22	Bretaro shaping a wooden door board using an axe. (Yakob, Duranmin, PNG) *Photo courtesy of Roger Lohmann.*	94
3.23	Lashed pole suspension bridge over the Fu River, Papua New Guinea (August 2005).	96
3.24	Belok abrading and smoothing a wooden doorboard with leaves (Yakob, Duranmin, PNG). *Photo courtesy of Roger Lohmann.*	97
3.25	(a) Wuniod using a marsupial rodent incisor tool to carve a wooden arrow tip; (b) close-up (Yacob, Duranmin, PNG). *Photo courtesy of Roger Lohmann.*	98
4.1	Fired clay types (adapted from Rice 1987: Table 1.2 and Hodges 1989[1976]: 53).	105
4.2	Vitrified pottery, forming a pottery "waster."	106
4.3	Generalized production process diagram for fired clay, focused on pottery (greatly simplified).	108
4.4	(a) Collecting clay and (b) processing dry clay by sorting out unwanted inclusions and breaking down lumps. (M. Nawaz and J.M. Kenoyer; Zaman; Harappa, Pakistan)	110
4.5	Types of turning tools for shaping and decorating clay vessels.	114
4.6	Zaman evening the base of a large vessel on a kick-wheel. The same kick-wheel is used as a tournette to add coils to the upper surface of large vessels whose bases are made in molds. (Harappa, Pakistan)	115

4.7	Nawaz painting the upper surface of a water jar on a kick-wheel used as a tournette. (Harappa, Pakistan)	116
4.8	Various firing structure types for firing pottery: (a) example of one type of ephemeral firing structure, an "open-air" firing structure; (b) example of one type of single-chamber firing structure, a pit kiln; (c) example of one type of multi-chamber firing structure, a double-chamber vertical or updraft kiln. (See Miller 1997 for full descriptions of these particular structures.)	122
4.9	Generalized production process diagram for vitreous silicates (greatly simplified).	131
4.10	(a) Glass melting debris and (b) glass molding debris. (Gor Khuttree, Peshawar, Pakistan) *Materials courtesy of the NWFP Directorate of Archaeology and Museums.*	139
4.11	Generalized production process diagram for copper and iron (greatly simplified).	146
4.12	Typical assemblage characteristics for non-ferrous metal processing. (Compiled from Craddock 1989:193, Fig. 8.2; Bayley 1985; Cooke and Nielsen 1978.)	155
4.13	Pouring molten copper alloy into bivalve sand mold. (Hyderabad, South India) *Photo courtesy of Lisa Ferin.*	160
4.14	Opening bivalve sand mold to reveal copper alloy bowl. (Hyderabad, South India) *Photo courtesy of Lisa Ferin.*	160
4.15	Blacksmith forging iron bar on expedient "anvil" in front of annealing fire. (Punjab, Pakistan)	163
4.16	Sinking and raising vessels from flat metal disks or sheets (a) flat metal disk; (b) sinking disk into wooden block with metal hammer; (c) bouging vessel over stake with wooden mallet; (d) raising vessel over stake with metal hammer; (e) planishing vessel over stake with metal hammer; (f) finished vessel. (Drawn after McCreight 1982.)	164
5.1	Map of Western Asia and adjacent regions.	169
5.2	Close-up of reed bundle construction, for Arabian Sea boat. (Redrawn after Vosmer 2000.)	171
5.3	Map of Southern California Chumash area, showing islands. (Redrawn after Arnold 2001.)	174
5.4	Sketch of Chumash reed boat. (Redrawn after Arnold 2001.)	174
5.5	Sketch of Chumash *tomal*, a sewn wooden plank boat or "plank canoe." (Redrawn after Arnold 2001.)	176
6.1	Map of Indus Valley region, showing sites of Harappa, Mohenjodaro, and Mehrgarh/Nausharo.	207

List of Figures xiii

6.2	Chronological systems for the Indus Valley Tradition, with approximate calibrated radiocarbon dates. (Modified after Shaffer 1992; Kenoyer and Miller 1999.)	208
6.3	Development of new talcose- and silicate-based materials (talcfaience complex) in the Indus Valley Tradition. (Data primarily from Mehrgarh-Nausharo studies by Barthélémy de Saizieu and Bouquillon—see text.)	211
6.4	Data sets used by archaeologists for the assessment of value.	213
6.5	Expected relative value diagram; this example uses Indus Integration Era object/material types.	215
6.6	Materials used to make red with white beads in the Indus Integration Era. ("Mature Harappan period")	219
6.7	Materials used to make white and blue beads in the Indus Integration Era. ("Mature Harappan period")	220
6.8	Map showing Pueblo regions and Casas Grandes Interaction Sphere. (Drawn after maps in Plog and Solometo 1997 and Walker 2001.)	229
7.1	Generalized production process diagram for copper and iron (greatly simplified).	243
7.2	Generalized production process diagram for fired clay, focused on pottery (greatly simplified).	244

DEDICATION

This book is dedicated to Jonathan Mark Kenoyer, for getting me started, and to Roger Ivar Lohmann, for helping me finish.

PREFACE AND ACKNOWLEDGEMENTS

This is a book that only a senior scholar should write, but only a junior scholar would be foolish enough to take on. I am grateful to Scott Bentley of Academic Press/Elsevier for waiting for a very junior scholar to become a slightly less junior scholar, through four changes of institutions, several states, and two countries.

This book is intended for a wide audience: for archaeologists who are interested in learning more about the study of technology; for technology specialists who are not archaeologists; and for anyone with an interest in the technical and social issues dealt with by past peoples in their processes of creation. It is unfortunately biased to Anglophone research, and particularly the large quantity of work from the North American tradition. There is, naturally, a tremendous amount of work on technology in other archaeological traditions and other languages, and I have tried to provide doors to some of these. There is a great deal of Francophone research of course, French being the language of a major stream of anthropological and archaeological technology research. My own research has also led me to considerable material in German and Italian, and I have tried to provide glimpses of these and other sources, primarily through English-language summaries. I hope readers will be encouraged to follow up these enormous pools of resources too seldom tapped by English speakers.

I have generally avoided two very large research sets on technology generated by archaeologists: the Classical Greek and Roman worlds, and Medieval to industrial period Europe. In both cases, there are quite a few general works on technological issues, engineering, or inventions, allowing points of access by the non-specialist. In addition, the relatively large amount of textual information available for these periods creates a very different research atmosphere, requiring additional primary tools of investigation to those discussed in this book. I have turned to these sources in a few cases where it was difficult to find such rich material elsewhere, as in my discussion of innovation and labor relations in Chapter 5, but for the most part readers should look to the rich literature within Classics, economic history, and history of science and technology.

This book owes its conception to my graduate advisor, Jonathan Mark Kenoyer, a lively advocate of trying to reproduce all sorts of technologies, an approach I have described as "exploratory experimental archaeology" in Chapter 2. It teaches one the process of research, gives insights into clues to look for archaeologically, and keeps the producers and consumers, the people and not the objects, firmly at the front of the picture. This hands-on approach to learning was extremely invigorating in combination with my earlier background in comparative analysis through time and across space, both in urbanism and in environmental archaeology. For the development of Chapters 3 and 4, I owe a truly great debt to Henry Hodges' *Artifacts*, with its clear and well-organized outlines of production methods worldwide for many different types of crafts, in a relatively short but surprisingly comprehensive volume. Hodges provides much more detailed information for European crafts (terminology in particular) than I have presented here—another reason I have not employed more European examples. I do not think I would have had the courage to write a book like this without this example. The structure of Chapters 5 and 6 of this book is modeled on Carla Sinopoli's *Approaches to Archaeological Ceramics*, since I have found her use of topical studies a wonderful teaching tool. Such a focus on issues related to people, not details of things, starts anthropology students off in the proper direction. The working title of this volume, *Archaeological Approaches to Technology*, is a grateful nod in her direction.

My own background is in South Asian archaeology, briefly as a paleoethnobotanist before being lured into the study of all the inorganic unidentifiable bits we were turning up, the remains of pyrotechnologies of various kinds. The site of Harappa, one of the great cities of the Indus civilization, was a good place to work, to be exposed to technology in many guises. I do not apologize for the preponderance of South Asian examples in this book—there is a tremendous amount of excellent work that should be more generally known. Perhaps the greatest influence that came from working with the Harappa team, in addition to the experimental and ethnographic work of Mark Kenoyer, was my introduction to Massimo Vidale in my very first season in 1990. In my first few weeks of learning how to deal with Old World large-scale urban excavations, I was "put in charge" of this already legendary senior scholar.

(One of the legends was that he constantly misplaced the half-dozen types of writing implements used by the Harappa project for maximum writing survival on different types of papers and plastic bags, so inventorying his equipment every morning was a major part of my job.) His kindness and patience, and general excellence as a teacher in matters from stratigraphy to working with craftspeople, and especially his sense of humor, made working with him a delight. Many of my best intellectual insights have been inspired by his ideas and innumerable publications.

Other members of the Harappa team besides Mark and Massimo have also had a great influence. Barbara Dales encouraged me to finish at a crucial point in my career, while Rita Wright has provided good advice and good examples through her knowledge of the literature. Richard Meadow has continued to encourage my sometimes bizarre perspective on agricultural systems, and human-plant relations more generally. The conservators and other archaeological students taught me a tremendous amount over the years, much of which has found its way into this volume. A special thanks goes to Muhammad Nawaz of Harappa, Pakistan, for all he taught me by example about excavation, pottery-making, and clay, as well as to all the crews I worked with there for their good humor and assistance, particularly my survey assistant, Saeed Ahmed Haderi.

This book as a book exists because of the enthusiasm of the students in the many technology classes I have taught, beginning as a teaching assistant to Mark Kenoyer at UW-Madison, where the art department graduate students sent me to the metals program and Eleanor Moty passed me on to special classes on African lost-wax casting with Max and Ruth Fröhlich at Haystack. While at Madison, my life was made livable by the other anthropology graduate students, who continue to contribute to my work: Lisa Ferin (who provided a number of illustrations for this text), William Belcher (especially for his gift of net-making materials), Rose Drees Kluth, Seetha Reddy, and Patti Trocki. James Knight and especially Kildo Choi made significant gifts of a number of difficult to find books on technology and archaeology.

My own technology course taught at the University of Michigan, with 35 bright and creative students investigating a variety of topics, was a great deal of fun. (I particularly appreciate their reticence about the near-fire in one of the labs during experimental attempts at making Japanese soot-based ink.) The Michigan faculty interest in the boxes of weird things I dragged through the halls, and reminiscences about their own experiences teaching hands-on classes in the past, has led in a number of unexpected directions, including John O'Shea's recommendation to Academic Press that a book like this would be useful, and that I should write it. Discussions with graduate students at Michigan have also had a significant impact on this volume, from continuing discussions with Kostalena Michelaki about bizarre slags, to Liz Sobel's gift of a box of obsidian pebbles and a demonstration of bipolar percussion on my office floor. Julie Solometo kindly shared a written copy of a conference paper and providing her MA paper, after my enthusiastic reception of her presentation at the 2000 Chacmool Conference at the University of Calgary. I hope this discussion of her work will encourage her to publish the full study, and continue with technology along with her current informative research into conflict. The graduate class on technology I taught at the Costen Institute of Archaeology, UCLA, formed the basis for the more theoretically-oriented graduate course I now regularly teach at Toronto. The hands-on part is the most interesting, though, and really far more informative for thinking about technology research, as I've found in my independent studies with a number of undergraduates at Toronto. At UCLA, students from various departments put together their diverse backgrounds to make for lively discussion and mutual growth, echoed by a similar graduate class a year later at the University of Toronto. To all these past students, if this book crosses any of your hands, I would love to hear what you are doing.

The Cotsen Institute of Archaeology is the place where this incipient book went completely off-track. Too many interesting things and people, and a staggering number of archaeological lectures by visitors from all over the world and right down the hall, opened doors to all sorts of new ideas and thinking. It has taken me years (literally) to come to grips with the tremendous variety of stimulations I received during my year as the Cotsen post-doctoral fellow in 2001–2002. May all researchers have a similarly distracting and stimulating experience at least once in their lives.

At all these institutions, at conferences, and by email, my peers have discussed, read, and commented on the various topics covered in this book—far too many to try to list here. Many, many thanks to all of you, and I hope the final version has not overlooked too many of your excellent suggestions along the way. I owe the greatest debt of all to the countless scholars, encountered in person or only through their writings, whose insights, data, and plain hard work have provided the background that made this book possible.

Finally, my family has always been supportive of my career choice and workload, showing particular tact in avoiding the whole topic of "the book" in family gatherings over the past few years. This book would never have gone to press without my parents' early teachings that being wrong is part of the learning process; I can only imagine how much I have likely gotten wrong in my determination to look outside my own areas of expertise for information. (But I'll take corrections!) Last but definitely not least, Roger Ivar Lohmann, my fellow graduate student, fellow anthropologist, and now husband and research partner, patiently read a thousand drafts, discussed everything from the definition of material culture to the typology of looms, offered advice and encouragement, and did far more than his share of the cooking and housework. He was particularly good at reminding me to discuss consumption as well as production, as I definitely have a tendency to focus on the latter. I want to thank him (I think) for his many difficult questions and requests for clarifications as this book was being written.

Heather Margaret-Louise Miller
Anthropology, University of Toronto

Almost every statement in prehistory should be qualified by the phrase: "On the evidence available today the balance of probability favours the view that." The reader is hereby requested to insert this or some similar reservation in most of my statements.

(Childe 1981 [1956]: 24)

CHAPTER 1

Introduction: Archaeological Approaches to Technology

... producers were necessarily interconnected through complex webs of interaction and inter-dependence.
(Sinopoli 2003: 6)

Each craftsman was dependent on his brother craftsmen of different trades in order to be able to carry out his own trade.
(Seymour 1984: 13)

The study of technology has always had a major role in the practice of archaeology. In all world areas and time periods, from every theoretical perspective, archaeologists have studied the ways in which people make things and interact with made things. Libraries of books have been written on topics including flintknapping, craft specialization, exchange networks, value, pottery classification, prestige goods, style, and artisans' roles in society. Given this volume of research, it is not surprising that most archaeological technology studies have focused on a particular technology, such as textile production or metal working. This is necessary for a deep understanding of particular technologies, given the complexity of the topics.

However, we do lose something by this necessary approach, and that is the ability to look at ancient technology from the perspective of multiple technologies. I was fortunate to be trained by a technology generalist, Jonathan Mark Kenoyer, who was sometimes disconcertingly adept at a variety of technologies. His assumption that one should investigate many technologies rather than being expert in one or two had a profound influence on my own choice of subject, the comparative study of Indus civilization pyrotechnologies. I have found such a broad perspective to be tremendously useful, if somewhat overwhelming. With this book, I hope to allow others to look for insights revealed by comparison between technologies, in combination with (not as a replacement for) the more usual focus on specific technologies.

Archaeologists tend to classify technologies into "crafts" or "industries" based at least partially on material or end-product type: clay vessel (pottery) production, metal working, basket making, stone object (lithics) production, woodworking, textile manufacture. This is in large part a result of having to classify objects without necessarily knowing their function, and is an essential part of the reality of archaeological research. Such groupings work very well for some cases, and are worth investigating from a theoretical as well as a practical perspective, as is shown in Chapters 3 and 4. But such material-based groupings can be counterproductive. Especially in cases with few textual records, it is rather rare for archaeological technology specialists to focus on technologies based on process or functional groupings such as transportation, luxury goods, communication, mining, or even agriculture (food production). This makes the transition within archaeology from collected data to social interpretation even more difficult. In addition, the general data-collection focus on groupings based on material or end-product as opposed to process or function is in great contrast to the focus of researchers in other fields, such as historians of technology, and can make discussion across disciplines rather difficult. Both such topic-focused studies and multi-craft comparative studies can help to ease such boundary crossings, by allowing movement away from a material-type focus and toward social interpretation as an ultimate goal, as I illustrate in Chapters 5 and 6.

This book is intended to allow archaeologists and others to obtain an introduction to production techniques and analytic approaches for many crafts or industries, and to show how archaeological specialists have used technology studies to address a wide variety of social questions about our ancestors. As this book is intended for a wide audience, I deliberately use as little discipline-specific terminology as possible, in order to accommodate readers from a wide variety of backgrounds, including specialists in different technologies. For example, pottery technology specialists and metal *technology* specialists use quite different vocabularies, obscuring those similarities that do exist between these crafts. Where technical terminology is necessary, I provide general definitions or make definitions clear from context, with references for further reading. I am thus using an explicitly comparative approach, and painting with very broad strokes. The references within the text are intended to provide entrances into more detailed and case-specific approaches, as well as a limited entry to the extensive literature on technology in languages other than English. I myself am an anthropological archaeologist, so am most familiar with literature from this field, but I have drawn on discussions by archaeologists in other disciplines as well.

TERMINOLOGY

Definitions can be used to set up terms of reference, and then employed in the investigation of a problem. This is how definitions are usually perceived and applied. However, defining a term can itself be part of the investigation of a problem. In the latter case, what a term means can only be revealed once we understand the full extent of the issue it symbolizes, and the process of creating a definition is how we go about investigating and solving an issue. I tend to follow this second practice. I use the process of defining something as a way to explore various facets of an issue or problem, so I am perfectly comfortable with shifting and negotiating terminology as I think through a topic. This is a method I frequently use in teaching of all sorts of archaeological topics, from political organization to technology. Therefore, I have no interest in using this volume to create a definitive account of how technology has been and should be studied in archaeology. Rather, I am far more interested in exploring the range of archaeological approaches to technology, particularly by organizing these approaches to allow comparisons across types of crafts.

We must have some common points of reference to begin, however. Dobres (2000: 47–95) offers a detailed discussion of the philosophical and historical definitions of "technology" and its roots, noting how diverse these definitions have been. In this book, I have instead employed a rough definition based on the way modern students of ancient technology tend to *use* the term, particularly within archaeology. The multiplicity of uses of the term technology has derived from the variety of approaches employed in the analysis of ancient technology. One line of this ancestry is the long-term interest in gesture and body posture during the production process, going back to Mauss and Leroi-Gourhan (Lemonnier 1992; van der Leeuw 1993). Another line is that of materials-based analyses by specialists in archaeology, the sciences and the arts, going back to Cyril Stanley Smith and others (Hodges 1989 [1976]; Kingery 1996, 2001). These lines overlap and interconnect; there is a long tradition of placing production within a social setting, of looking at the people making and using the pots that archaeologists find. For example, in *Artifacts*, one of the few archaeological single-volume works covering a multiplicity of technologies, Hodges (1989 [1976], originally 1964) distinguished technology from the study of stylistic details of artifacts. By this, he implied that technology was about the *process* of production rather than the endpoint (objects). Subsequent scholars such as Schiffer and Skibo (1987) and Lemonnier (1986; 1992), studying technology from quite different perspectives, noted as components of technology both the actual manipulation of physical objects and the shared (or secret) human knowledge involved. Merrill (1977:vi; also 1968) explicitly referred to technology as "the culture surrounding the actions or activities involved in making or doing things."

Researchers in a variety of disciplines have increasingly stressed the nature of technology as *practice*, as ways of doing or making something, of organizing work and people into systems involving new words and new mindsets as well as new tools (e.g., Lechtman 1977; Lechtman and Steinberg 1979; U. Franklin 1992; Basalla 1990; Lemonnier 1992, 1993; Pfaffenberger 1992; Kingery 1993, 1996; Dobres and Hoffman 1995, 1999; McCray 1998; Schiffer 2001). That is, we add not only hands but also faces and minds to our study of the way things are made. To go even farther, Franklin (1992) specifically discussed technology as ways of *doing* something rather than simply ways of making something (that is, creating an object), so that there are technologies of prayer and of storytelling as well as of pottery production and weaving. This returns us to Leroi-Gourhan's view of dance as a technology; gesture as well as material culture is important. In short, the archaeological study of technology employs a decidedly holistic approach. Kingery (1993) notes that for archaeologists and art historians, the term technology generally brings to mind the production process, whereas historians and philosophers of technology usually think of the design process. This difference is in large part based on their respective materials of study, which in the past traditionally included differences in data sources (object vs. text), scales of production and distribution (single objects or small-scale vs. large-scale production), and to some degree time periods (pre-historic vs. historic). Increasingly, members of these disciplines have bridged these differences over the past few decades. Albeit still in progress and not without its problems, the resulting examination of the entire process of technology, including the social context, is one of the most exciting trends in the study of technology.

As this is a book about many approaches, I am rather broad in my definition of basic terms, rather than trying to give a more narrow definition that may not fit some of the thematic studies discussed. Of necessity, there will be some fluidity in my use of terms. So to begin with basics, what do I mean by *technology*? I think of technology in the context of an outwardly expanding, nested set of actions and relationships: from production itself, to the organization of the production process, to the entire cultural system of processes and practices associated with production and consumption. "Technology" is commonly used to refer to each of these sets, perhaps because the study of ancient technology has developed from so many different perspectives, or perhaps because the everyday use of the term technology also has a wide variety of meanings, as consultation of any dictionary will indicate. To clarify my discussions,

Introduction: Archaeological Approaches to Technology

I will define terms to distinguish between these nested sets of actions and relationships, although the boundaries between these sets are blurred.

Production is the actual process of fabrication or creation, including both the material objects involved and the techniques or gestures used. Rice's (1996b: 173) "manufacturing" and Costin's (1998b: 3–4) "crafting" are roughly equivalent terms, but I have chosen production to allow consistency with the enormous literature on the "organization of production." Furthermore, although both manufacturing and crafting can be explicitly defined and used as referring to the creation of objects from start to finish, I find that manufacturing often conjures up an image of factory-style workers participating in a segment of an object's production, while crafting tends to provoke an image of individual artisans designing and creating an object from start to finish. Although the image of "production" is often closer to that of manufacturing, I incorporate both of these images in my use of production. The term technology itself is often used to mean the same thing as my term production, both in everyday speech and by specialists, particularly in reference to techniques of fabrication or production. The discussions of the sequence of stages by which objects are produced (production sequence or *chaîne opératoire*), as presented in Chapters 3 and 4, relate to both production and the organization of production as well as the life-histories of objects.

I define *organization of production* as the organizational arrangement within which production takes place. This may refer to one artisan working on an object from start to finish, or it may refer to a system of specialist workers, managers, and materials procurers. The many discussions of craft specialization in the archaeological literature falls within in this category. Although the social and economic aspects of the organization system are usually emphasized, ideological attitudes from politics or religion can also have a major impact on the organization of production. The term technology is also sometimes used to refer to the organization of production; for example, Franklin (1992: 18–20) distinguishes between holistic technologies and prescriptive technologies as methods of organization of production that parallel the distinction between crafting and manufacturing noted for production above.

In this book, I use *technological systems* or just *technology* primarily in the broadest sense, to refer to the active system of interconnections between people and objects during the creation of an object, its distribution, and to some extent its use and disposal. In other words, technology or technological systems can be roughly described as the processes and practices associated with production and consumption, from design to discard. Consumption (including distribution, use, and disposal) is thus included in a technological system, as it should be, given the strong links between production and consumption on a number of levels as illustrated throughout this volume. This characterization of a technological system is a useful shorthand for a complex term, if not a precise definition. It derives primarily from more complete discussions by Franklin (1992), Kingery (1993; 2001), and Lemonnier (1992). Pfaffenburger's (1992) "sociotechnic system" is similar to my use of the term technological system. While many archaeologists would apply production, organization of production, and technology strictly to the creation of physical objects, I favor Franklin's (1992), Walker's (2001), and Sinopoli's (2003) suggestions that these terms can also refer to the creation of non-objects, such as music, dance, rituals, and poetry.

In contrast with the process-oriented focus of technology, I use the term *material culture* to refer to the interactions between people and objects (usually finished products). This includes both the ways in which people perceive objects and react to their culturally prescribed meanings, as well as the ways in which people give meaning to objects. As in the short-hand definition for technology, this is at best a rough approximation of the uses of a complex term. It has the added advantage of probably satisfying no specialist in the topic, so I do not privilege one current mode of thought over another. Readers can explore more precise (and contradictory) definitions in a number of recent books and collections (e.g., D. Miller 1985; Lubar and Kingery 1993; Kingery 1996; Chilton 1999; Glassie 1999; Schiffer and Miller 1999), and the recently-established *Journal of Material Culture*. The use of the term material culture has often been linked strongly to discussions of technology, as in Lechtman and Merrill's (1977) edited volume, *Material Culture: Styles, Organization, and Dynamics of Technology*. Although definitions and degrees of linkage have varied, almost all summary volumes on crafts or materials, such as those referenced in Chapters 3 and 4, employ case studies related both to material culture studies and technology studies.

Archaeologists, although hotly debating its exact meaning, are generally comfortable with the use of the term material culture. Perhaps in large part this is because it is our starting point—ultimately, our work comes down to dealing with the associations between surviving "things" and past people. However, other scholars, especially other anthropologists, still have reservations about the term, and a short exploration of this issue is revealing. For one thing, the term "culture" itself has been one of the most debated terms in anthropology—defining technology is easy in comparison. The very use of this term thus rouses shades of disquiet. Beyond this, material culture is sometimes seen as implying a potentially sloppy equation of culture with objects, something archaeologists have worried about for decades. It is therefore particularly important to emphasize that material culture is not the same thing as "objects." Material culture is about *interactions* between people and things, and especially about information encoded in things. This is a sensible perspective for those who define culture as information learned and transmitted to others, consciously or unconsciously, which is at least a portion of most definitions of culture. If culture is contained in information stored in human memories and passed on to others, then information stored in written documents and conveyed into human memories must also be seen as culture. Artifacts of all types also encode information, which can similarly be conveyed to human memories. From this we can describe material culture as the information encoded in and expressed by human use of objects. Whether the meaning conveyed to others is the same as the original meaning intended by the maker or user is a problem, of course, but this is the case for all forms of information communication. Note that objects may be used to simultaneously record *and* express cultural information, a point of much discussion in the literature on material culture, as well as the archaeological literature on style, as discussed in Chapter 5. This point also pertains to a more general discussion about defining culture as information. One part of the debate about culture refers to the existence of culture in two forms: an unexpressed form as a (mental or physical) record of information at the individual level, and an expressed form as objects, behaviors, or discourse at both the individual and group levels. Materializing culture, or the process of creating material objects (that is, technology), is one way that shareable, learned information is expressed, parallel to communicating information through speech and behavior.

So for the purposes of this book, I will use the following terminology. Both technology studies and material culture studies use the investigation of "things" to understand past and present societies. Material culture studies tend to focus on the interactions between people and (finished) objects, while technology studies tend to focus on the human practices and processes associated with (object) production. However, this distinction is blurred in application as researchers often study both processes and finished objects, particularly in examining the life histories of objects. Studies of the way people think and speak about production and the organization of production also link the study of technology and the study of material culture. If not the same entity, the two are yin and yang—one cannot be understood without an awareness of the other, as Lemonnier has stressed (1992: 2–3).

ARCHAEOLOGY AND TECHNOLOGY STUDIES

What does the archaeological study of technology have to contribute to the broader study of technology and material culture? The most obvious contribution of archaeology is that of a broad perspective, which can either follow a particular society through time or range across many societies. It provides information about the development and acceptance of new objects and new production techniques, and about changes in past economies, social structures, and political organizations in relation to the invention or adoption of technologies. While my examples in this book refer to time periods from the Paleolithic to the present, I have not implied any sort of evolutionary development. A discussion of the evidence for and against a general increase in technological complexity around the world through time would be a book in itself. Rather, my aim is to show specialists in other fields, such as the history of technology, communication, women's studies, studio art, and sociocultural anthropology, that archaeologists have created a variety of examples well worth their time to investigate.

Archaeologists also have a great deal to offer other disciplines in their development of methodologies and theoretical approaches, both for teasing information out of objects and for looking at societies in their entirety. Archaeologists are obliged by one of their primary techniques, excavation, to deal with societies for the most part as a complex whole, rather than in separate packages of ritual beliefs, economic units, or centers of political power. Whether participating in rescue operations or conducting normal excavations, archaeologists constantly face the loss of the past through destruction, including the oft-cited ethical dilemma of necessarily destroying a site in the process of excavating it. Therefore, the practical field reality requires archaeological projects to recover as broad a range of data as possible, no matter the particular goals of the project, in order to maintain an ethical standard of work. Whether such recovery methods are practicable in most cases, the methodological *ideal* is still one of complete holism.

In addition, archaeologists have reconstructed many of the ancient processes of production, from manufacturing techniques to labor organization. These reconstructions are of use for modern artists, craftspeople, labor specialists, and managers, as they portray the strengths and weaknesses—both technical and social—of different pathways to the production of objects. Economic historians might benefit from archaeological reconstructions of economic competition and its effects on past societies, and the pivotal roles that ancient technologies sometimes played in the distribution of power within and between social groups, affecting social status and political structure. Finally, most people in today's world can use a reminder that new technological inventions themselves have seldom altered society. Rather, it is the ways in which the society or individuals within it use and adopt new technologies that result in social change. This is a topic I develop in further in Chapter 5.

I have indicated what archaeological approaches have to offer students of technology in other fields. But what does the study of technology have to offer archaeology as a discipline? In the best cases, technology studies build bridges between scholars in different locations, between different disciplines, and between different traditions or approaches. The study of technology in archaeology has been outstandingly international, with the intersection of researchers from different countries working in different world areas and time periods. The use of archaeometric analysis has fostered collaboration with scholars in the sciences, the focus on objects has encouraged interaction with colleagues in the arts, and the importance of technology in both the prehistoric and the historic periods has provided links with researchers in the historical disciplines. Technology studies increasingly cross the divisions between the sciences, social sciences, and humanities.

This integrative role of technology studies in archaeology makes it difficult to disentangle and label separate traditions or schools of research. (See the introductions to Hamilton (1996) and McCray (1998) for two different examples of such summaries.) Many of the studies discussed in this book draw on multiple traditions of thought and method, even for the basic reconstruction of production sequences. These include such apparently disparate traditions as the *chaîne opératoire* approach (Inizan, et al. 1992), the history of technology and engineering design (Kingery 1993, 2001), operations process management (Bleed 1991), and the use of practice theory (Dobres and Hoffman 1995). Individual researchers use different combinations of approaches, depending on what is useful for the question involved. Recent approaches employ aspects of materialism, where economics and environment are seen as the most important factors in the nature of social groups, and idealism, where idea-based sectors like religion, ideology, and kinship are favored as the major factors in social behavior and change. Of course, some or even most researchers may defend one approach as far superior to others. In my opinion, the best research privileges no single approach, but considers the applicability of several. In the following chapters, my discussion will illustrate different theoretical and methodological approaches. Although I do occasionally discuss problems and shortcomings of particular techniques and studies, it is easy to find critiques in the literature. I have chosen instead to focus on the creative ways archaeologists have negotiated around these shortcomings.

OVERVIEW OF VOLUME

In this chapter, I have introduced the terms production, organization of production, technology, and material culture. I have explored the usefulness of an archaeological approach for technology studies, and the importance of technology studies for archaeology. This book is unusual in my examination of multiple technologies. It is difficult to fit the discussion of even one set of technologies, such as metal working or transportation, into a single volume. But the study of multiple technologies is necessary if we are going to examine commonalities between different technologies and how past people perceived, exploited, and supported them. Like the humans who create them, no technology exists in isolation from others. In showing how technology studies help us to understand past societies, this book provides an entry point into the rich body of archaeological research examining the interplay between human groups and their technologies. In addition to describing how archaeologists study ancient technologies, I will concentrate on answering archaeological questions about past societies through technology studies.

The next chapter covers the methods that archaeologists use in their study of past technologies. I will touch on field techniques; various approaches to artifact examination; experimental studies; and the use of ethnoarchaeology, ethnography, and historical accounts. This will be a rapid tour of a large corpus of information, giving brief examples of archaeological studies from a diversity of regions and time periods. Entire texts are written on each of these methods, and I have provided initial references for further reading. As for all types of archaeological topics, the range of methodological tools is vast and impressive; archaeology employs perhaps the most diverse range of techniques and methodologies of any discipline. This is not surprising, given that archaeology is essentially the study of all aspects of human life in the past. Techniques and methodological approaches can be applied from all disciplines relating to modern humans, from descriptions of their physical environment to analyses of their political structure and philosophical viewpoints. On top of this must be added methods employed in extracting information from the bits of ancient garbage we find, requiring skills in computer use, statistical analysis, and analytical research techniques.

Chapters 3 and 4 provide overviews of the primary production processes for a number of material or end-product categories, including the production of flaked and ground stone; fibers, basketry, and textiles; sculpted organics such as wood, bone, and shell; pottery and other clay-based ceramics; faiences and glasses; and metals. These general overviews of production employ the traditional material-based archaeological approach to technology, and include directions to numerous texts focusing on each of these technologies. My goal here is to provide readers with enough background to understand the basic structure of production for these major crafts. I also provide brief archaeological illustrations of production organization or product consumption from a variety of times and places, to give some glimpses into the variations possible in these aspects of technology. Diversity in production techniques, organization of production, and the whole system of technology is further explored for the thematic studies provided in Chapters 5 and 6.

The thematic studies are my favorite part of the book, for they show the vistas that archaeology has opened for us on the creativity of ancient people, and also their failures. We see their clever manipulations of local resources and their struggles with environmental conditions. We view the variety of technological innovations, with new techniques adopted and new techniques rejected based on economic conditions, social systems, and traditions of practice.

Throughout, there is a sense of the intricacy of craftspeople's roles, and that of their products, in economic, political, social, and ritual practices. The thematic studies are organized by process or functional topic rather than material type, including labor organization, economic exchange, value and status markers, religious rituals, technological style, environmental considerations, and consumption demands. Some studies are primarily economic in focus, often concentrating on production as a material base for economic power. Other studies adopt idealist perspectives to examine the relation between technology and social identity. Readers may skip to these chapters first, if they desire, but a good basic background in both the ancient production processes and the archaeological methods of analysis will make the nuances of these thematic studies more apparent. The first thematic study, on reed-bundle boat technology in the Arabian Sea and Southern California, serves not only as an illustration of exchange and wealth accumulation, but also as a model case to illustrate the use of archaeological finds, laboratory analyses, ethnography, and experimental studies. These data are employed to provide information on production techniques and processes, as well as on the role of these boats in their very different societies. Subsequent thematic studies have been chosen to fit the topics I present, and to provide wide coverage of different world areas and types of societies. The examples are thus a mixture of classic, well-known studies, and lesser-known research, often still in progress. I only wish I could include more of the studies I found in researching this book, as the range of archaeological ingenuity and tenacity involved is impressive.

In the final chapter, I am able to draw on all these thematic studies and descriptions of production processes to present an idea of the sort of framework that might be used to view ancient (and modern) technologies comparatively. Inspired by W. David Kingery (1993), I have gone beyond the examination of production processes to think about frameworks for comparing crafts throughout their entire technological process, which includes social desires, contexts of use, and discard features. For example, I examine different types of production processes to determine the degree to which storable, transportable, potentially multi-purposed semi-finished products such as metal ingots, lithic blanks, or thread for cloth can be produced within different crafts, and then discuss how these differences can have important impacts on the organization of production for these crafts, as well as the potential for flexible response to supply and demand.

Ultimately, this book places the many technological revolutions of recent times (that is, the past few hundred years) in context with the development of new technologies in earlier periods, particularly with respect to their social context. The development of plastics has interesting social parallels to and differences from the development of other new materials in earlier time periods, such as the Old World faiences described in Chapter 6, as these materials were used first for ornaments and luxury containers. As discussed in Chapter 5, inventions may be known, used, and abandoned, only to be rediscovered in somewhat different form and widely adopted centuries later, when the economic and social conditions are more appropriate. The place of women in the work force changed dramatically on more than one occasion prior to the twentieth century upheavals. By looking at the long time scales provided for us by archaeological research, we can begin to place our own times in perspective, and perhaps take a broader look at how technological changes of the past century relate to the equally astounding social changes in work force, material culture, and daily life.

CHAPTER 2

Methodology: Archaeological Approaches to the Study of Technology

You know my method. It is founded upon the observation of trifles.

Sherlock Holmes (Doyle 1988: 214)

Archaeology is a supremely challenging, puzzle-solving activity. This is a large part of the attraction for its practitioners, and the reason why so many people are fascinated with the profession. The buried, seemingly inaccessible past has a powerful pull on our human curiosity. The ways in which archaeologists make the past speak through fragments of refuse is one part of the mystique, our modern-day shamans loosing the tongues of our ancestors. More prosaically, the process of archaeological investigation has considerable pedagogical value. The practice of archaeology provides an extremely useful way to illustrate the broadest techniques of problem solving and procedures of research, incorporating as it does sciences, humanities, arts, social sciences, and practical application, all in one discipline.

This chapter outlines the methodologies employed in the archaeological investigation of past technologies, methodologies that informed the thematic studies in Chapters 5 and 6, and the other examples throughout this book. How do archaeologists interested in technology (or other subjects) know what they know about the past? To be an archaeologist is to be a generalist, as one never quite knows what will turn up on a field project. Some of the methods of investigation used for the archaeological study of past technologies were developed in archaeology, such as excavation and surface survey, but many are borrowed or adapted from a wide range of other disciplines. As in archaeological research in general, technological studies often employ collaboration with specialists in other fields, including conservators, art historians, chemists, material scientists, architects, engineers, botanists, geologists, miners, artists, and other modern craftspeople.

Like the imaginative detective work of Sherlock Holmes, archaeological methods involve long and tedious studies of minutiae, extensive collections of odd reference materials, and personal immersion in a person's work and life. As with Holmes' analyses, the investigation of trifling minutiae can yield startling results, although the most exciting cases for the practitioner are not always those of greatest interest to the public. For example, when asked about "my most exciting find," I usually mention the inscribed rim of an ancient Harappan stone vessel, face up in a muddy surface scattered with pottery sherds and other debris. Or an abandoned meal, a pit full of shellfish buried in the back of an urban alley-way some 4500 years ago. These *were* exciting moments, when something very rare or something very human seemed to stretch out a hand in greeting from the past. However, the *most* exciting find I have ever made was not an event, but a growing realization over months of fieldwork in 1993, when I spent my afternoon off-time happily sorting several thousand kilograms of burned clay bits (Figure 2.1). (This was a proceeding that also raised the spirits of the nearby pottery sorters, since at least *their* job, the sorting of tens of thousands of plainware sherds, involved objects with recognizable shapes.) The excitement for me in this process was the dawning recognition that only a few, easily identifiable fragments of these thousands

FIGURE 2.1 Sorted fired clay fragments at the Indus urban site of Harappa. The only fragments indicative of high-temperature firing are in the small pile to the far right.

of clay bits were diagnostic of high-temperature firing, and could be used to accurately locate buried and eroded craft activity areas across the surface of the large urban site of Harappa. Why this might be of interest to anyone other than an obsessed pyrotechnologist is something I will come back to later in this chapter.

In this chapter, I have somewhat artificially divided my discussion of archaeological methods into sections about archaeological field techniques; artifact examination; the organization of data; experimental archaeology; and ethnographic, historical, and ethnoarchaeological sources. In reality, there is a great deal of interaction between all of these methodologies. The boundaries between them are blurred, and other researchers might divide the study of technology in different ways. In addition, research in one area affects research in others, so the order in which these sections are presented is not necessarily the order in which archaeologists conduct research. The way artifacts are recovered (field methodology) affects their examination and analysis, and the examination of artifacts influences further fieldwork. The results of experimental archaeology and ethnoarchaeology affect the examination of artifacts by providing new traces to look for and new interpretations of their possible production and use. Finally, the type of methodology used at every stage is affected by the researcher's theoretical approach, the types of questions that the researcher is asking about the past.

Archaeologists studying different time periods, world areas, and scales of societies use various combinations of techniques, depending on what is appropriate for their questions about the past, for the societies under study, for the environment of the region, and for their operating budgets. Technology is a part of this past, so although most production and use areas are fortuitous finds, archaeologists generally do their best to record the information they recognize. However, if a specialist in the particular craft involved is not present on an excavation or survey project, it is easy to miss clues—there is simply too much to know. One aim of this book is thus to provide a general overview to technology studies for a range of crafts, so that every archaeologist has a better chance at maximal recovery of important data.

In the best studies of ancient technologies, researchers use as many appropriate methods as possible to thoroughly study the problems dealt with by ancient craftspeople, and the solutions that were found. Such problems were often technical. How to achieve the temperatures needed to fire a pot? How to make a colorfast dye? What tool to use for a tricky bit of carving? When to irrigate a field? But ancient craftspeople also faced economic and social issues, such as fluctuations in supplies and in demand for their products, difficulties in the recruitment of competent apprentices, issues with workshop organization and control of production, and the negotiation of their standing within their societies. Any investigation of past technologies has to recognize that these sorts of economic, social, and political issues were just as problematical for ancient craftspeople as the technical challenges they faced. Investigations of technological change and innovation must include such factors facing producers, as well as similar issues facing their consumers, to understand the motivations for new development and the reasons for the adoption of new technological products. As in all other types of archaeological investigation, the archaeological study of technology must place objects in their social context, and remember that we are interested in people, not only their things.

ARCHAEOLOGICAL FIELD TECHNIQUES: DISCOVERY/RECOVERY

The archaeological field techniques used to investigate past technologies are the same as those used by archaeologists to investigate all aspects of the past. Archaeological field techniques are traditionally divided into two types or phases of study: survey and excavation. Other sources of data for the investigation of technologies include documentary and oral sources, which are discussed below, as well as the examination of objects and records in collections and archives. This last source of data is often particularly important for technological studies, as discussed in the section on Examination of Artifacts.

Most field projects identify and record materials related to technologies, especially artifacts from the production process. Such artifacts include *tools*, such as spindle whorls for spinning thread; *by-products*, such as pottery wasters, or cores and debris from stone working; and *installations*, such as oil presses or metal furnaces. For some crafts, such as stone working, caches of *raw materials* can be used to identify storage or production areas. High densities of finished *products* have also been identified as possibly representing storage or consumption areas, an important aspect of the overall technological system. For example, such high densities of products have been used to assess pottery production and consumption; Feinman (1986: 355) describes the combined use of kilns, misfired sherds, and "abnormally dense concentrations" of specific types of finished pottery vessel fragments to identify pottery production areas in surveys of the Valley of Oaxaca, Mexico.

In most environments, the only surface finds preserved are those made from the sturdiest materials, particularly stone and baked clay. So there is considerable bias in the crafts that can be studied directly, favoring crafts with nonperishable products, tools, by-products, or installations. This is also the case for buried materials, although there is often slightly better preservation so that bone and metal objects remain. Only in the "best" environments from the extreme dryness of Egypt and Peru to the bogs of northern Europe and swamps of Florida, do we typically recover objects made from wood, fibers, feathers, and other organic materials, and these are almost always from excavations, not survey.

SURVEY

Survey can be used as a prelude to excavation or as an end in itself. It is no exaggeration to say that the development of survey techniques has revolutionized archaeology in the past century, allowing us to see humans within a larger landscape rather than focusing on a settlement or two. The study of regional interactions and changes, through survey, are now more common archaeological foci than the perspective from a single settlement. Regional (intersite) survey is also used as a means to select areas or settlements for further investigation, usually by excavation. Of course, humans in the past frequently saw their world from the perspective of one site, their home settlement, so it is important for archaeologists to do site-focused research as well. Like regional survey, survey across a single settlement (intrasite survey) is used to pinpoint areas within a site for excavation. But as with regional survey, intrasite survey in itself is also an important method of archaeological analysis for questions of site patterning and function (Hietala 1984; Kroll and Price 1991).

The literature on archaeological survey is vast, ranging from introductory surveys of methodologies to edited volumes of results. Considerable thought and ink has been devoted to improving the representativeness of survey work; that is, the degree to which the artifacts or sites recovered reflect the actual distribution of past activities across the landscape. Banning (2002) provides discussion of these issues and further references; the *Journal of Field Archaeology* is another useful source for specialist discussion. For a more general introduction, most of the introductory field texts cited in the Excavation section below discuss the practice of survey as part of general archaeological fieldwork.

The most basic archaeological survey consists of a team of people walking across a landscape and recording the artifacts and features they find (Figure 2.2). There are considerable refinements to this technique, specialized for particular landscapes, types of remains, and questions under consideration. For example, where ancient landscapes are buried under shallow alluvium, as in the Midwest of North America, surveys frequently employ "shovel testing," digging a small test pit at set intervals along a walking trajectory to test for buried artifacts. Another technique is to take soil cores using an auger to investigate the nature of subsurface deposits; this is a good way to determine the extent of buried shell mounds or midden deposits. Geophysical prospection techniques such as gradiometry and resistivity are also employed to check for buried features like storage pits, kilns, stone walls or defensive ditches,

FIGURE 2.2 Surface survey, with flags used to mark objects until they are mapped, recorded, and/or collected.

while more remote sensing techniques from airplanes and satellites use imaging with various light spectra to see whole settlements and other large-scale patterns on the landscape (A. Clark 1990; Schmidt 2002; Scollar, et al. 1990; Weymouth and Huggins 1985; Zickgraf 1999).

Materials collected from general archaeological surveys have served as the basis for innumerable studies of technology, examining production, distribution, and consumption. As a case in point, the general surface surveys in the 1960s across the Basin of Mexico around the ancient city of Teotihuacan have provided data for a number of subsequent technologically-oriented studies, conducted long after the survey areas were swallowed up by modern development (Cowgill 2003: 38–40; Millon 1973; Millon, et al. 1973). These include studies of obsidian, lapidary, ceramic, and ground stone production, distribution and consumption, which have provided social as well as technical information about these crafts. For example, Biskowski (1997) documented differences in cooking technologies between different socioeconomic classes at Teotihuacan, based on analyses of the ground stone collections and archived data about their find-site distributions from these early surveys.

In addition, there are increasing numbers of surveys focused directly on technological issues, as shown in the thematic studies and other examples in this book. Surveys have been used to investigate changing product distribution patterns over time, for information about consumers and economic trade networks. Technologically-oriented surveys have also contributed significantly to our understanding of cases where control of production has—and has not—operated as a source of political power. Archaeologists often examine the exploitation of particular sources of raw materials, through regional surveys coupled with provenance (or provenience) studies, to determine where people acquired their stone or clay or metal ore, and how this related to changing patterns of regional trade and communication.

For example, some of the most elaborate field research on ancient technologies has incorporated survey as well as excavation to study ancient metal-producing regions, focusing usually on large-scale mining and smelting operations. Craddock (1995) describes many examples of such work, as do articles in the journal *Historical Metallurgy* and the *Institute for Archaeo-Metallurgical Studies (IAMS) News*. In these integrated, multidisciplinary studies, visual and remote sensing survey work is usually just the first step, and is followed by exploration of mines, excavation of production areas and settlements, and a variety of artifact studies and specialist investigations. Aspects of copper mining and smelting are further discussed in Chapter 4. However, the mining and smelting stages of metal working, and the initial reduction stages of stone working, are rather unusual in the visibility of their by-products, especially for large-scale and long-term use of a raw material source. Other stages of production and other crafts are more elusive. Rather than being near a raw material source, their production areas may be less predictably located within a landscape. For most stages of most crafts, production areas are likely to be in or near settlements. Such production areas are often found fortuitously, during the course of regular excavations at a settlement. Sometimes enough remains from production are visible on the surface to allow for planned excavation of a craft production area, but this is not as common as fortuitous discovery. Improving the intentional location of production areas using survey has been among the goals of most intersite and intrasite surveys concentrating on craft production.

EXCAVATION

Excavation involves the uncovering of objects and other traces to provide clues to past activities and beliefs, ultimately leading to reconstructions of past events and ways of life. As with survey, the range of techniques for excavation is enormous, and varies on the basis of the questions being asked, the type of site, the local environment, and the amount of time and funding available (Figure 2.3). The literature on excavation is extensive, and the best entry point for non-specialists are introductory texts designed for archaeological field schools or field methods courses (e.g., Drewett 1999; Hester, *et al.* 1997; McMillion 1991; Roskams 2001). As with survey, there has been considerable concern about the representativeness of excavation areas, as excavation of entire sites is not possible for any but the smallest settlements or campsites.

FIGURE 2.3 Excavation; shovel-skimming the base of a trench in a plowed field. *Photo courtesy of Patrick Lubinski.*

Indeed, if possible, archaeologists prefer to leave a proportion of a site unexcavated, so that it may be excavated in the future with new and better methods. Unfortunately, with the rapid expansion of modern populations, only a few sites can be protected, and these are seldom the "average" village or short-term camp. A large part of archaeological excavation is thus concerned with the ethical issue of recognizing and recording as much information as possible about all aspects of the past at each location.

Increasing the field identification and recovery of craft working areas is one of the greatest challenges currently facing the study of ancient technology. As noted above, many major crafts leave few remains under most conditions of preservation. This is especially true of most stages of the processing of organic materials like textiles, basketry, hide, wood, and food, but it is also the case for particular stages of the processing of inorganic materials, such as the forming stages of clay object production, or fabrication stages of metal object manufacturing. Identification of the production areas for such crafts or production stages are dependant upon the recognition of a few preserved tools, such as roughly shaped clay or stone loom weights or smooth stones used for polishing pottery. These tools were few in proportion to the number of objects produced, were rarely discarded or lost, and in many cases are seldom identified in the field at the time of discovery. Quite often these production tools are only recognized long afterwards when specialists examine a collection, if these nondescript objects have been retained.

In contrast to the more usual case of fortuitous finds of production areas during excavation, many of the studies described in this book derive from research projects devoted to technological issues from their inception, either as the main focus of an entire team, or the focus of one or two members of a team. These studies are thus not typical examples of archaeological investigations, but represent concerted efforts to collect information about specific technologies. Besides collecting products, tools, and production debris, such directed projects sometimes make systematic studies of soil, vegetation, or trace element samples, especially for the location of more ephemeral production areas. These areas are identified through traces of remains from hide tanning, fabric dyeing, metal production, food and drink production, the production of various building materials, and so forth, as described in articles in journals such as *Archaeometry, Geoarchaeology,* and *Environmental Archaeology* (formerly *Circeae*), and other publications (e.g., Murphy and Wiltshire 2003).

Distribution and consumption patterns of produced items are also typically studied by specialists long after the excavation work has been completed, in part because of the extensive laboratory analysis often involved. Astonishing information about the level of specialized production and distribution in the ancient Mediterranean world, including the use and re-use of widely-traded ceramic containers such as amphora, has come from finds from underwater archaeological investigations of shipwrecks, as well as finds from land excavations and textual accounts (Bass 1975). In short, methods of discovery and recovery in archaeological technology studies are as diverse as the technologies under study.

THE EXAMINATION OF ARCHAEOLOGICAL REMAINS

I have divided my discussion of the examination of archaeological remains into "simple" and "complex" methods. Simple refers to the simplicity of the tools needed (the human eye, perhaps aided by a basic microscope), not the simplicity of these analyses, which usually involve a great deal of training and experience. It is easy to overlook the tremendous amount of information provided by these methods of simple examination, in the excitement over the results generated by studies using complex, expensive equipment. Both approaches are needed for thorough studies of ancient technology. Often simple examinations are essential for allowing study of a large range of materials, to allow the proper choice of the handful of samples that can then be tested using complex examination methods. For most technologies, getting samples of the full range of debris or tool or product types can be essential to the success of a project, as numerous projects working on metal production have particularly stressed (Bachmann 1982; Craddock 1995; Tylecote 1987). This point is an important one, because a key aspect of archaeological research is the analysis of remains within their chronological and spatial context. Samples tested at random, no matter how well preserved, tell us much less than samples placed within a context.

The examination of objects includes their long-term preservation, so that they may continue to be studied by future scholars with new techniques or new ideas. The field of conservation includes the cleaning, stabilization, preservation, and in many cases the reassembling or restoration of objects prior to long-term storage (Pye 2001; Sease 1994). With these goals, conservators are often major participants in the study of technology. In the course of treatment of objects prior to storage, conservators often make significant observations relating to production. As materials specialists who need to know the chemical and physical attributes of objects before they can properly treat them, conservators are especially qualified to ask and answer questions about the production process. In addition, many conservators do extensive research on technologies, including experimental as well as analytical tests, because the way something was made influences the choice of processes for conserving it. Conservators, like archaeologists, use a variety of methods to examine archaeological remains, as seen in *Studies in Conservation*, *Reviews in Conservation*, and other publications of the International Institute for Conservation of Historic and Artistic Works.

The conservation of objects related to production and use is especially important so that such objects can be available for additional research in the future. I have stressed how often nondescript objects that are key to understanding production processes are unrecognized or misclassified, until subsequent studies of collections are done by specialists or until new methods of analysis are developed. In both cases, this can take place decades after the original surveys or excavations, as illustrated by the studies of materials from the Teotihuacan surveys mentioned above. The preservation of field records in archives is thus equally important. The analysis of remains within their chronological and spatial contexts is a key aspect of archaeological research, and objects that cannot be placed in context provide significantly less information about the past. Collections management and ethics are receiving increasing attention in archaeology, as the dilemmas of funding and preservation are multiplying (Childs 2004; Sullivan and Childs 2003).

SIMPLE VISUAL EXAMINATION AND MEASUREMENT

The simple examination of archaeological objects by eye or by low-powered magnification allows researchers to record aspects of the object such as form, color, and elements of design, as well as simple physical measurements such as dimensions, angles, and weight. Such basic systematic recordings are still the

Methodology: Archaeological Approaches to the Study of Technology

cornerstone of archaeological typologies, providing the essential codings of morphology and design elements for objects and architecture. These records of morphology and design—the shape of stone tools, the size of bricks, the decoration of column plinths or ivory pendants—provide the data for cutting-edge studies of style and for century-old techniques of chronology building. For the student of ancient technology, changes in objects over time can provide valuable insights into changes in production techniques or the organization of workshops, or changes in supply networks for raw materials that might reflect political turmoil. Archaeology shares many of these techniques with art, art history, and other disciplines interested in stylistic analysis, for which there is a vast and contentious literature. Examples relating to such analyses are discussed in Chapter 5 in the thematic study on style and technology.

Simple visual examination and measurement of objects is also the primary method of identifying natural objects directly or indirectly related to human actions. Remains from animals (including humans) and plants are identified by direct visual means, through comparison with collections of bones, teeth, shells, and scales from known animals (Klein and Cruz-Uribe 1984; O'Connor 2000; Reitz and Wing 1999), or seeds, wood, tubers, pollen, and phytoliths from botanically identified plants (Pearsall 1989; Piperno 1988). With regard to technology, such remains can provide information about food procurement techniques, food processing, leather and fur production, the use of woven mats, bone object manufacturing, or the location of areas where plant fibers were stored or processed, as in many of the studies cited in Chapter 3.

Reconstruction of past production techniques and processes is also often based on simple visual examination. No study of technology should neglect this step, where surface scratches can reveal hafting techniques for stone objects (Martin 1999: 96–107); uneven joins or particular types of cracks can provide clues to pottery manufacturing techniques (Rye 1981); and sizes of spindle whorls might reveal the type or fineness of thread being spun (E. W. Barber 1994; Teague 1998). Simple visual investigation of waste materials such as stone flakes or vitrified clay fragments can provide information on the techniques of production that occurred at a location.

For example, fired clay fragments recovered from sites of the Indus civilization have been extremely useful in determining the location of production areas, and even the structure of firing installations. Work by teams at the 4,500-year-old cities of Mohenjo-daro and Harappa have shown that only a few specific types of melted clay fragments are pieces of high-temperature pottery kilns, copper-melting furnaces, or faience production tools. As my long afternoons of sorting heaps of debris at Harappa revealed (Figure 2.1), the vast majority of fired and melted clay fragments came from over-fired clay "nodules" probably used for a range of functions from foundation gravel to heat-retention in pottery kilns (Figure 2.4). Lower fired clay fragments come

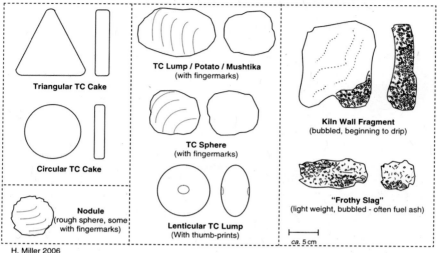

FIGURE 2.4 Fired clay object types from Harappa, including debris from high-temperature production, nodules, and lower-fired clay cakes and balls.

from flattened cakes and balls in a variety of shapes and sizes, the majority of which seem to have been used for heat retention or pot-props in domestic cooking hearths (H. M.-L. Miller 1999: 154–167). The excitement of these findings came from the realization that not all types of fired clay are equally appropriate for providing information about past technologies. By sorting and identifying these nondescript materials more accurately, it has been possible to locate potential high-temperature production sites more accurately, by looking for the comparatively rare types of fired clay fragments indicative of firing installations (H. M.-L. Miller 2000). This is particularly valuable for survey in populated areas or at tourist sites like Harappa, as such uninteresting fragments were almost never collected by either local visitors or past archaeologists, and so remain relatively undisturbed for modern surveys. Being able to locate at least some production sites from surface survey not only allowed us to plan excavations of craft areas, but also to begin to examine the way crafts were organized within the settlements, giving clues to social, economic, political, and perhaps ideological aspects of production (H. M.-L. Miller 2000, 2006; Pracchia, et al. 1985; Tosi and Vidale 1990).

Even more specific studies of these unprepossessing objects have provided extremely useful information about the shape and functioning of long-destroyed kilns. Pracchia and Vidale (Pracchia 1987; see also Pracchia and Vidale 1990) used the types and orientations of melted and dripped clay to

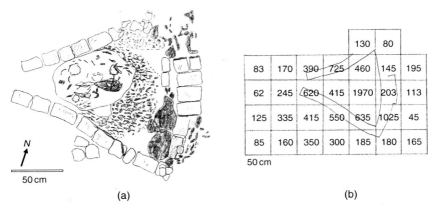

FIGURE 2.5 (a) Pottery kiln remains from Mohenjo-daro, in relation to (b) number and distribution of drips. (Redrawn after Pracchia 1987.)

determine the shape and structure of a destroyed pottery kiln at the ancient city of Mohenjo-daro (Figure 2.5). From the shape of some fragments and the pattern of melted drops on them, they reasoned that the kiln was double-chambered, with a lower chamber for the fuel and an upper chamber where the pottery was fired away from the smoke of the fuel. (See Chapter 4 for more details on pottery firing.) From the locations of the most vitrified fragments, they determined the orientation of the kiln and the position of flues for encouraging the draft (increasing the airflow) and raising the heat of the kiln. All of these conclusions were based solely on simple examination of objects in the field and their spatial distributions.

COMPLEX EXAMINATION OF PHYSICAL STRUCTURE AND COMPOSITION

The examination of artifacts also involves more complex methods of measuring physical structure, as well as composition. Again, "complex" does not imply that such studies require more knowledge than "simple" examination, but rather that the tools involved are more complex, and usually much more expensive. Such studies, often referred to as *archaeometry* or *archaeological science*, are carried out in both the field and in the laboratory, and borrow techniques and approaches from physics, chemistry, material science, engineering, geology, and other disciplines. There is an enormous and growing variety of analytical techniques, destructive and nondestructive, quantitative and semiquantitative, focused on composition and on structure. English-language overviews include Jakes (2002), Leute (1987), Sciuti (1996), and Tite's (1972) classic older text, while Goffer's (1996) multi-language dictionary of terminology is also very useful, given the multilingual nature of archaeometry. Rice (2000) edited a recent volume on integrating archaeometry into anthropologically-oriented archaeology. More detailed studies can be found in journals such as *Archaeometry, Journal of Archaeological Science, Archaeomaterials*, and the various conservation publications given above, as well as the proceedings of the Archaeological Sciences Conference and of the International Symposium on Archaeometry.

Archaeometric techniques of analysis are diverse, employing almost every type of equipment for visualizing and recording materials, shapes, colors, and other attributes of objects. The physical structure of objects is examined using various types of visual microscopy and different wavelengths of light, to reveal shape, crystalline structure, and also composition. Other compositional analyses make use of traditional wet chemical methods, or employ bombardment with atomic particles, as in the various types of neutron activation analysis (NAA). Different techniques are used to identify the presence of different elements or compounds in a sample. In addition, different methods of analysis can provide quantitative, semiquantitative, or qualitative results. Quantitative results will give the actual amount of particular elements or compounds present in a sample, semiquantitative results provide relative amounts of elements or compounds, while qualitative results simply indicate whether an element or compound is present or absent. It should thus be clear that different techniques are useful for different kinds of questions, so that knowing the best method or methods to use to answer particular questions is a fundamental aspect of archaeological science. The use of multiple methods of analysis to examine the same material or object allows researchers to collect complementary data, or to reinforce conclusions.

The usual image of these archaeometric methods of analysis is that of a few precious artifacts ensconced in a high-tech laboratory, studied by a white-coated expert. This is certainly an accurate picture of high-tech labs around the world dedicated to the analysis of ancient artifacts, from the alloys of metals (Scott 1991) to the types of paint (articles in *Techne* Issue 2, 1995) used in the prehistoric and historic past. An example of how the laboratory study of products can provide evidence for the reconstruction of economic and technological systems is Galaty's (1999) analyses of broken pottery vessels, especially drinking cups, from the Mycenean palace at Pylos. Galaty used petrography, the visual study of the minerals in the pottery, to identify the different places where the drinking cups were made, and ICP spectroscopy to determine the existence of two separate systems of pottery production and distribution within this small ancient state.

But archaeometric research is also carried out beyond the walls of laboratories, with some or all stages of the process occurring at find spots or raw material sources, by scruffy-looking experts squatting on the ground sorting through piles of ancient garbage. Complex methods of analysis also include methods for the discovery and analysis of production areas, in the field or through subsequent lab work, including geophysical prospection, soils analysis, and trace element analysis, as discussed in the Field Techniques section above and in journals such as *Journal of Field Archaeology, Geoarchaeology, Archaeological Prospection, Prospezioni Archeologiche Quaderni,* and *Environmental Archaeology.* As specialists, these researchers might work with several teams, acting as consultants for particular issues, or they might concentrate on sites of special interest, such as ancient mining operations. The lab specialist and the field specialist often need to be the same person, or at least people in close contact, to ensure the correct recovery of representative samples for both simple and complex methods of artifact examination. The choice of materials submitted for analysis will determine the appropriateness of the laboratory analyses for answering archaeological questions. At this point, methods from the sciences, from radiocarbon dating to chemical analysis, are so completely interwoven into the practice of archaeology that it is difficult to determine what exactly qualifies as a separate subfield of archaeometry. Instead, in the best cases, these methods for studying the lives of past humans have become one of the many different specialties held by archaeologists who contribute as part of a team.

ORDERING AND ANALYZING DATA

All of the methods of archaeological data collection described here produce large quantities of information, information that must be catalogued and ordered to allow people to make sense of it. The ordering of data can be based on any criterion, such as statistical analyses based on quantitative measurements, or the visual sorting of objects using qualitative observations seen in the example of Harappan fired clay fragments above. Archaeological systems of classification are often divided into those based on characteristics relating to technology, style, or morphology, the last sometimes mistakenly referred to as "functional typologies" in spite of the fact that shape does not always relate to use. Most actual systems of classification employ a combination of such characteristics.

Whatever the method of ordering, the resulting structures or systematics (Banning 2000) form the basis for pattern recognition studies used to make or test suggestions about the past. The nature of these classifications has generated tremendous debate in archaeology. As I have noted, some of the most contentious issues of sampling and survey methodology relate to statistical issues of representativeness (Banning 2002). This concern with sample representativeness extends to all aspects of archaeology, especially but not exclusively when using statistical methods of analysis (Banning 2000; Carr 1985; Drennan 1996; Shennan 1997; see Sharer and Ashmore 2003 for an introductory overview). Other debates have centered around the meaning of classifications, whether these categories are all necessarily "artificial" or "etic" categories imposed by archaeologists, or whether it is also possible to discover "natural" or "emic" categories—that is, categories employed by the people who actually made or used these objects. Sharer and Ashmore (2003: 296–298) and articles in Whallon and Brown (1982) provide overviews of this debate, particularly as it applies to pottery. Every analytical topic in archaeology, from counting animals to sorting stone flakes, has its own methods of classification and its own debates about them.

The archaeological examination of artifacts always involves contextual or relational data, primarily the analysis of spatial or chronological distributions. Analysis of the distribution of artifacts both chronologically and spatially is standard practice in archaeology, including the distribution of materials related to technology. The chronological ordering of archaeological data is the key to the study of change, involving both innovation and adoption. New methods of dating often allow refined accuracy, but the older methods of stratigraphic and other forms of relative dating are still the foundation for most archaeological discussions of change, including technological change. Spatial distributions have become especially important now that survey is such a key aspect of archaeology, and archaeologists have been among the major users and developers of geographic information systems (GIS) (Lock 2000; Wheatley and Gillings 2002).

These object-focused analyses form one scale of analysis. Other scales of analysis are analyses focused on stages of production, and analyses of crafts within an entire economic and social system. As much as possible, archaeologists must consider all scales of analysis as they decide how to order their data, since they are all affected by these choices. Nor are these scales of analysis hierarchical in importance; as with molecular-, organism-, and systems-oriented biology, different scales of analysis are appropriate for different questions. As is the case in biology, all levels must be strongly researched, or the research will draw false conclusions and overlook important aspects of ancient technologies. Questions about social relations are popular in the archaeological study of technology at present, as the thematic studies in Chapters 5 and 6 show, just as experimental studies were popular in the 1970 and 1980s. However, these approaches are only as valid as the object-focused data sets on which they draw, just as object-focused studies must be designed with attention to the types of questions about economic or social systems they will be used to explore.

RECONSTRUCTING PRODUCTION PROCESSES; CHAÎNE OPÉRATOIRE

For technological studies, a common bridge between the ordering of data and the interpretation of past actions is the use of production process or production sequence diagrams. These diagrams outline the stages in the process of production, and often the raw materials used and the products and by-products produced at each stage as well (e.g., Figure 2.6). The focus of

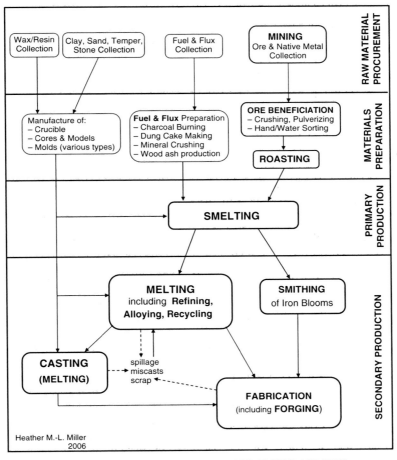

FIGURE 2.6 Example of a generalized production process diagram for copper and iron (greatly simplified). Explained further in Chapter 4.

production process studies is on the alterations to the materials, resulting in the creation of a finished product. The object is the center of attention. Such diagrams are widely used by archaeologists as a methodological tool to clarify their reconstructions of the stages of production, as has been done in Chapters 3 and 4.

The *chaîne opératoire* approach employs diagrams often identical in appearance to production process reconstructions, and is considered by many researchers to be a similar type of production process study. However, other researchers believe the *chaîne opératoire* approach is fundamentally different in conception, noting that in this approach, as originally defined by Leroi-Gourhan, the focus of the production sequence reconstruction is not only on the alterations to the materials, but also on the gestures—the hand and body movements—used in the alteration of materials (see Inizan, *et al.* 1992 for an introductory discussion). The producer, the person producing the object, is the center of attention, although in archaeological cases their movements are necessarily inferred from the alterations to materials. This description fits well with Chazan's (1997: 723) definition of *chaîne opératoire* as "the unfolding of a technical act." Lemonnier (1992: 26) defined it as the "series of operations involved in any transformation of matter (including our own body) by human beings." Developed for the study of Paleolithic tools, and widely used in lithic analysis at present, the *chaîne opératoire* approach is increasingly applied to other crafts, from pottery (van der Leeuw 1993) to basketry (Wendrich 1999).

Both of these data analysis tools, production process diagrams and the *chaîne opératoire* approach, infer stages of production from archaeological data using analogies based on the types of studies described in the next two sections: experimental, and ethnographic and historical investigations.

ANALOGY AND SOCIOCULTURAL INTERPRETATION

Production techniques and styles are only a portion of technological studies. Technology also includes information about the specialized knowledge and organization of the people making things. For example, are there specialized basket makers, or can most people in the society make a basket—or both? Does one person make a basket from start to finish, or are a number of people needed, each of whom has a particular task, such as cutting the reeds or fibers and processing them, or preparing and applying dyes, or designing and making the basket itself? Technology, as I have defined it, also includes use and discard patterns. Who uses these baskets—only near kin, or people living in the same village, or a far-flung network of consumers? Is the actual

Methodology: Archaeological Approaches to the Study of Technology

object being exchanged the basket itself, or the contents of the basket? Are the baskets used only for very specific purposes, or can they be used in various ways? How can we know all of this for the past?

Archaeologists use many sources of evidence to move from the incomplete, fractured data sets we recover in field, laboratory, and collections research, to the interpretation of past ways of life, including technological systems. I have briefly discussed the use of archaeological data itself to suggest or test hypotheses about the past. Researchers also draw on knowledge accumulated from other sources, outside of the actual archaeological data. Particularly in anthropological archaeology, which seldom has textual or oral accounts of the period under investigation, analogical reasoning is the primary approach used to move from the results of all of these methods to reconstructions of past societies.

To give a very simplified example, we might find from excavations that a small settlement, Site A, has no hearths inside houses, but only in the shared central area between the houses (Figure 2.7). We might try to explain the reasons for this pattern by looking for similar cases in ethnographically or historically known societies living in similar settlements. If we find that for the vast majority of known cases, hearths inside houses are used to cook food primarily for the people who live in that house, while outside hearths are used to cook foods shared by the entire group, we might use analogical reasoning to suggest that the people of Site A typically shared their food between households. That is, we could use observed similarities in data about

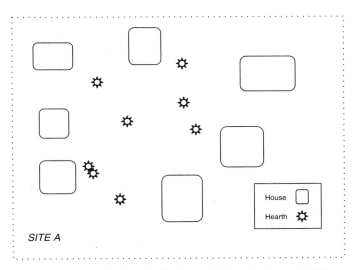

FIGURE 2.7 Example of an idealized site with hearths between buildings.

the location of hearths in both the archaeological and ethnographic cases, to suggest similar reconstructions about food cooking and sharing practices for the archaeological case.

The use of analogy to explain past situations requires careful attention to the range of possible explanations, whether these alternative explanations are based on ethnographic or historical cases, experimental studies, or deduction. Using the example above, an alternative explanation might be that since Site A is located in a hot, dry climate, people would not want fires inside their homes. Another alternative could be that inside cooking hearths did exist, but were archaeologically invisible because they were built off the ground inside metal or clay braziers which were removed when the houses were abandoned, while the outside hearths were used only for special functions like seasonal feasts, or processing large amounts of foods for storage, or even for firing low-fired pottery. A third alternative might be that uncooked food was considered to be dangerous, dirty, or polluting, and could not be brought into sleeping and living areas. To support any one explanation, the archaeologist needs to provide additional supporting lines of evidence for the explanation and/or provide evidence to discount alternative suggestions.

Most analogical reasoning is formal, based on the assumption that if things have some similar attributes, they will share other similar attributes. Relational analogies, in contrast, are based on inherent or causal linkages between the attributes of the cases. However, relational analogies are all too often limited to linkages based on the physical properties of materials. For example, we can be fairly confident in our use of ethnographic or experimental examples of grain crop processing to interpret archaeological assemblages, as there are physical laws governing the behavior of the sorting process that limit the number of ways in which large quantities of grain can be successfully separated from chaff (Hillman 1984; Jones 1987; Reddy 1997). Flaking of chert has a similarly limited number of ways in which flakes can be removed from core stones, allowing reasonable reconstructions of the flaking process for archaeological finds. Focusing on the physical properties of materials makes such relational analogies particularly useful for the study of production processes. Cunningham (2003) notes that there is no inherent reason why relational analogies cannot be extended beyond causal processes based on physical properties. However, the larger number of alternatives and subsequent complexity of the studies needed often makes such research more difficult and time-consuming, requiring long-term dedication and support.

Weak or strong, explanation by both formal and relational analogy is necessarily the primary method of archaeological explanation. It is thus not surprising that debates in the 1970s and 1980s about the appropriateness of analogical reasoning for archaeology shook anthropological archaeology in North America to its foundations (see Cunningham 2003 for an excellent summary;

Gould and Watson 1982; Hodder 1982; Wylie 1982, 1985; Yellen 1977). These disputes about the uses of analogy converged with debates about the very nature of knowledge about the past, from the degree to which it is ever possible to know the "real" past, to the relative importance of materialist and idealist factors in the process of social change. Of course, philosophers, psychologists, and physicists debate the degree to which we can know even the "real" present, so there has been plenty of room for discussion. At the same time, archaeologists began to explore (and argue about) the usefulness of a wide variety of approaches to understanding the past, often lumped together as "processual" or "post-processual" approaches to archaeological theory. For brief, initial introductions to these developments and further references, see Sharer and Ashmore (2003, especially Chapters 3, 13, and 17) and Renfrew and Bahn (2000, especially Chapters 1 and 12).

This period of debate in archaeology is particularly relevant to the archaeological study of technology. Many of the debates about the use of analogy included examples of the reconstruction of ancient technologies, as techniques of production provide most of the best examples of relational or causal analogies. However, analogical reasoning was and is used to explore "technology" in all aspects of the definition presented here: in terms of techniques of production of objects, the social organization of the production process, and entire sociotechnological systems. On the one hand, technology—in the sense of techniques of production—was seen as one of the things archaeologists could definitely know about the "real" past. At the same time, archaeologists began to highlight social and ideological aspects of production as ignored yet essential aspects of past technologies, as exemplified by many of the articles in seminal edited volumes promoting gendered (Gero and Conkey 1991) and Marxist (Spriggs 1984) approaches.

Since the 1990s, most archaeologists seem to have settled into an acceptance of a pluralistic discipline, using theoretical approaches as diverse as the methodological approaches employed. However, while the terminology in use varies extensively, there is still an insistence on ensuring that both theoretical and methodological approaches are relevant for the questions and site conditions at hand, and that all interpretations are subjected to critical assessment. Archaeologists are quintessentially students of the material, and the ethical requirement that collected objects and their contexts be recorded, analyzed, published, and curated acts as a central anchor for the field. Perhaps because of this grounding, the pendulum of fashion in theory and method can only swing so far within the discipline as a whole.

Experimental studies and ethnographic or textual sources are the two major sources of analogies for the interpretation of technology studies, as for many other aspects of archaeology. These two aspects of archaeological investigation of the past, centering as they do on the application of analogies, have been subject to considerable criticism as a result of the debates in archaeology alluded to above. They nevertheless continue to be of central importance in archaeological research into all aspects of ancient technology.

EXPERIMENTAL ARCHAEOLOGY

Experimental archaeology takes many forms and is widely practiced, so that studies have been published in many languages and journals (*Journal of Field Archaeology, Journal of Archaeological Science,* and *Lithic Technology,* to name just a few). Mathieu (2002) provides a useful summary of the Anglophone literature on the definitions and purposes of experimental archaeology, including summaries by Ascher, Coles, Ingersoll, Yellen and Macdonald, and Schiffer and Skibo. Entry into the strong Francophone tradition in experimental archaeology can be accessed in English as well as French through works by Inizan *et al.* (1992) and Lemonnier (1992, 1993), articles in the journal *Techniques et culture,* and the work of researchers in Centre Nationale de la Recherche Scientifique (CNRS) research groups and universities highlighted on their informative websites.

Mathieu (2002: 1–2) draws on this literature to define the practice of experimental archaeology as the use of controllable, imitative experiments to replicate and/or simulate past objects, materials, processes, behaviors, or even entire social systems (although he admits the last are primarily within the purview of ethnoarchaeology, as discussed below). The purpose of these replications, in Mathieu's definition, is to allow the researcher to create and test hypotheses that can be used for archaeological interpretation through the generation of analogies. Mathieu emphasizes four aspects of experimental archaeology through this definition: (1) the controlled, repeatable nature of the experiments; (2) the fact that these are re-creations, and so not necessarily the same as the original objects or behaviors in the past, even if successful; (3) the use of these experiments to create and test specific hypotheses about the past; and (4) that the final goal of the process is the generation of analogies (preferably relational analogies) used to interpret the traces of past processes or actions.

Less formal studies are also useful, what I would call "exploratory" experimental archaeology (also see Mathieu 2002: 7). Initial attempts to make and fire pottery, flake stone tools, or reproduce textiles, even if neither informed by archaeological finds nor strictly measured and monitored, can be very useful beginnings to the more formal process of experimental studies or ethnoarchaeological analysis. Informal studies, by being less structured, less time-consuming, and generally less expensive, allow more room to explore possible alternative production processes, with their concurrently higher risk of failure. Even "playing" with production processes helps archaeologists to adopt the perspective of the producer rather than the observer of the object, and so acts as a very useful pedagogical tool. The conclusions from such informal studies must be further verified with archaeological checks and formal experimentation, of course, but exploratory experimentation serves an important purpose in the analysis of past technologies. Although not so controlled, it contributes to the goals of experimental archaeology: the replication of past objects or techniques to test the feasibility of proposed reconstructions of production or use techniques, and/or to create and test new analogies.

Experimental replications illustrate gaps in knowledge and design flaws not envisioned until the actual construction of the object or execution of the process is attempted. An essential aspect of replication for an *archaeological* project is the constant checking between experimental reconstructions and the archaeological materials, in a cycle of research, reconstruction and comparison. This is one of the central objections to many replication projects, because simply to reconstruct a plausible and workable reconstruction is no guarantee that this was the way the object was made in the past. For example, as discussed in Chapter 5, Heyerdahl (1980) used construction techniques derived from modern reed boats made both in Western Asia and in South America to create and sail a reed boat in the Arabian Sea. However, Cleuziou and Tosi (1994) subsequently determined from archaeological bitumen finds from ancient reed boats that Heyerdahl's boat, while an effective seagoing craft, was not constructed in the way reed boats were actually built in the ancient Arabian Sea. Thus, as an exploratory experiment, Heyerdahl's work was very useful, but comparisons with subsequently discovered archaeological remains showed that further replications needed to be changed to match the actual Western Asian boats.

Experimental archaeology can be especially useful for creating and testing hypotheses about past production and use activities, given its emphasis on controlled conditions. Knecht (1997) tested the different raw materials used for Upper Paleolithic projectile points in Europe, to gather new insights about the design and relative performance of these objects. She compared the production processes, use characteristics, and ease of repair for antler, bone, and stone projectile points used for hunting. By testing a range of raw material types, rather than only stone, Knecht was able to make a broader suite of conclusions about the possible choices involved in the use of particular raw materials.

Experimental archaeology is also used to generate analogies for reconstruction of past production *processes*, as seen in Vidale's (1995) examination of the by-products generated at different stages of groove-and-snap production of steatite disk beads. By examining a large number of archaeological finds and attempting various processes of production experimentally, Vidale was able to propose a likely sequence of production for these beads. He was also able to recognize cases of alternative production sequences, which opened the door for suggestions about the nature of the craftspeople working at the site.

Although these two examples are from recent work, experimental research focused primarily on the re-creation of objects and the technical aspects of production are no longer as prominent in the archaeological literature as they were in the 1970s and 1980s. In the late 1990s and early 2000s, technology studies became more visibly focused on social aspects of production, including organization of production and the status of producers (e.g., Costin and Wright 1998; Dobres and Hoffman 1995, 1999; Hruby and Flad 2006; Schiffer 2001). There are notable exceptions to this generalization, of course, but overall this is a positive sign of archaeologists' renewed commitment to a focus on people rather than things. However, as the two recent examples show, reconstructions of production processes should not be scorned, as we must have a solid understanding of production to proceed to our social interests. We still simply do not know how some materials and objects were created or used. After all, one of the great fascinations of archaeology is the ingenuity of ancient people. Some of the best records of that ingenuity can be found in the process of production, whether it is the production of vegetables, cities, or artificial stone.

ETHNOGRAPHY, ETHNOARCHAEOLOGY, AND HISTORICAL ACCOUNTS

There are many accounts of production, organization of production, and technological systems found in oral histories and folklore, and in ancient and more recent written accounts. These accounts can be of great help to researchers in a variety of fields, including archaeology, in the reconstruction of past technologies. However, each source of information has problems of its own. Nothing quite matches the actual demonstration of production by living people, but this is no longer an option for most areas of the world and most crafts. Interviews with past specialists are often the best that can be hoped for, and perhaps experimental recreations of their past work. Other sorts of oral histories and folklore are usually only obliquely concerned with technology. Much of the written information was noted by travelers or by nonspecialists, with varying degrees of detail and accuracy. Some written accounts were collected by ethnologists in the course of their research in order to preserve some vestiges of changing ways of life; others were written by historians or encyclopedists, to preserve the achievements of their day. A very few accounts were written by specialists in the technology as manuals. Within archaeology, historical and Classical archaeologists make the greatest use of written accounts, and are most familiar with the special challenges such records present (R. J. Barber 1994; Bowkett, et al. 2001; Orser 2004; Weitzman 1980). As fragmentary snapshots recorded for differing purposes, historical and oral accounts are intentionally or unintentionally biased, and far too often written by observers who were not practitioners and so missed or misinterpreted essential aspects of production or nuances of social relationships. Care must be taken in the employment of such accounts when reconstructing past technologies.

Socio-cultural anthropologists have had varying degrees of interest in the topic of technology over the past century. Details of technologies were routinely recorded in early ethnographies, especially in "salvage" ethnographies focused on recording rapidly changing ways of life. The popularity of cultural ecological approaches in the 1950s and 1960s encouraged the study of technologies, particularly those associated with food production. Although less common in more recent times, excellent accounts of technologies still continued to be produced, found in such diverse ethnographic sources as the Shire Ethnography series, the CNRS journal *Techniques et culture*, and individual monographs like MacKenzie's (1991) detailed account of net bag production in Papua New Guinea, which provides specifics of the production processes as well as nuances of the interactions between production, use, and gender relations. An ethnographic account of ground stone adze production in Papua New Guinea is given in Chapter 3, and illustrates the range of information important to the producer—not only production technique, but also potential value as trade items, the social network of people who were allowed to access the raw materials directly, and the network of groups involved in exchange.

As shown in Chapter 5, in the discussion of wooden plank boat building by the Chumash of Southern California, ethnographic information can provide analogies for all aspects of technology. Ethnographic accounts not only provide models for the production of wooden plank boats in terms of the techniques and materials employed, but also models for the organization of the production process, and the broader role of this new technology in the society. In the case of the Chumash, ownership of these boats was primarily limited to those who could afford to commission their construction. Such ownership resulted in increased wealth and social status, and may have played a pivotal role in the rise to power of regional chiefs (J. E. Arnold 1995, 2001). This ethnographic example provides a model of *one possible way* that wealth and power might be gathered and monopolized by members of a social system through the use of a new technology, resulting in changes in social structure. Other social effects of new boat technologies are also possible, such as in New Guinea, where boat ownership was not so restricted (J. E. Arnold 1995). These ethnographic and historical examples provide alternative models for archaeologists to test for other cases of past maritime societies for whom we have no ethnographic or historical information. The application of ethnographic or historic information to archaeological problems, involves analogical reasoning, and as with experimental conclusions, these data must be compared with the archaeological remains, in the cycle of continuous checking described above.

Ethnologists are rarely interested in discard patterns, however, and so many archaeologists have turned to *ethnoarchaeology*, studying living people in order to gain the insights they needed to make analogies with the past, particularly with regard to material culture and technology. David and Kramer's (2001) recent book, a fitting climax to the long careers of both authors in ethnoarchaeology, provides alternative definitions of the term, as well as its history and many case studies. Nicholas David's massive web-based bibliography of ethnoarchaeological articles is also an important source for those looking for information on ethnoarchaeological research.

David and Kramer (2001) and Cunningham (2003) both stress the central role of analogous reasoning for ethnoarchaeology, as do all of the ethnoarchaeological studies discussed throughout this volume. They also highlight the diversity of topics and theoretical approaches associated with ethnoarchaeological projects. As with experimental archaeology, ethnoarchaeology is used to generate and test analogies for use in interpreting archaeological remains. And as with experimental projects, alternative explanations must always be considered since ethnoarchaeological examples are not necessarily the same as the original objects or behaviors in the past, even if the material traces produced seem identical. Ethnoarchaeology has been a particularly rich source of alternative explanations and new analogies, which as Cunningham (2003: 404) notes is something to be celebrated rather than scorned about this approach.

Ethnographic and ethnoarchaeological information can be very effectively combined with experimental archaeology to provide precise models for archaeological testing—or for further experimental or ethnoarchaeological tests. A large percentage of technologically focused ethnoarchaeological studies are of this type, asking specialists in a craft to remember the "old ways" of doing things, or to try to make an ancient object somewhat different in shape or material than the modern objects he or she usually produces. A well-known example of the resurrection of long unpracticed technologies is the extensive research in Africa on iron smelting undertaken by a number of teams, as summarized in Childs and Killick (1993). Experimentation with ancient shapes by modern specialists in bead-making has been undertaken during research in Khambhat (Cambay), Gujarat, India, as discussed in more detail in the Stone section of Chapter 3 (Kenoyer, *et al.* 1994). In these sorts of studies, archaeologists work with modern craftspeople who have the skills to attempt various possible replications of ancient techniques, even if these are not the techniques the modern craftspeople typically use in their own work.

Another well-known example of the power of such combined experimental and ethnoarchaeological approaches is the research that has been done into the phenomenal ocean navigating abilities of ancient Polynesian sailors. The oral accounts of tremendous journeys across the Pacific Ocean have been corroborated by creating experimental replications of boats and voyages, using a combination of oral and historic accounts, material remains and depictions, and most importantly expert traditional advice from various parts of Polynesia and Micronesia (Finney 1998). This information has been used to reinterpret historic objects, such as objects now known to be ocean map models, as well as to strengthen the archaeological evidence for the navigational abilities of Polynesians in the past. It has also led to renewed vigor in modern Polynesian interest and pride in traditional navigation, sailing, song-making, and other crafts, creating new traditions as well as reviving the old (Finney 2003).

All of the methodologies and sources of information discussed in this chapter have informed archaeological investigations of past technologies. In the next two chapters, I outline basic production processes for most of the major material classes studied by archaeologists, including stone, fibers, wood and bone, fired clay, faience and glass, and metals. Since archaeologists usually sort objects by material class for analysis, at least initially, I have used material type as my primary classification principle for the discussion of production. This aids in my goal of providing basic information about the process of production for a number of crafts, to allow archaeologists to recognize and recover clues to technology in the field or lab. Chapters 5 and 6 employ different methods of organization, around topics rather than material classes. Both of these approaches aid in the comparative examination of multiple technologies, as discussed in Chapter 7.

CHAPTER 3

Extractive-Reductive Crafts

> The video displays the world according to basketry: the baskets are the protagonists, the basket makers are the supporting actors... [while] the book focuses not only on the baskets, but also on the production process and the basket makers.
>
> (Wendrich 1999:2)

One of the most fascinating aspects of the study of ancient technology is the variety of ways in which people have created objects and living environments. Chapters 3 and 4 provide the basic grounding needed to facilitate the exploration of this variety. In the next two chapters, I outline the production processes of six material-based classes of ancient crafts: stone, fiber, wood and other sculpted organics, metal, fired clay, and faience and glass. I have selected a combination of those crafts most widely used by ancient people and those most often studied by archaeologists. In the latter case, there is a strong focus on crafts involving less perishable materials.

Each section of Chapters 3 and 4 provides a brief outline of the overall production process for each craft complex, to provide readers with a basic introductory framework. I have focused on the essentials of the most common production techniques, and provided references for more detailed overviews of the diversity of production processes. These referenced texts provide good introductions to more detailed research and case-specific publications, as well as older general studies. For broad coverage of a wide range of crafts, Hodges (1989 [1976]) is one of the few single-volume texts available, and is still essential although somewhat outdated. Older multivolume works, such as the landmark compilations of Forbes (1964–1972) and Singer et al. (1954–1978), are very useful as long as more recent studies are also consulted to check for updated knowledge. The Shire Ethnography and Shire Archaeology series are outstanding resources for a wide variety of regionally-specific craft production techniques, especially the many crafts employing organic materials such as woodworking, textiles, and basketry. Henderson (2000) provides recent coverage of the study of inorganic material groups (glass, ceramics, metals, and stone) that is more detailed than my brief synopsis, particularly for analytic methods.

Production is only one part of technology; organization of production, consumption patterns, and the entire technological system in its human context are all part of most archaeological definitions of technology as discussed in Chapter 1. Nonetheless, as Wendrich (1999) has comprehensively illustrated in *The World According to Basketry* book and video quoted above, it is necessary to understand these various production methods, at least in outline, in order to understand the requirements and rhythms of the production process. This knowledge of the production process allows us to identify points in the process where alternatives exist in techniques of production, and so where producers might have made different choices based on a range of factors from the available materials to cultural traditions. For example, flat metal objects such as arrow or spear points can be made in a number of ways. One way would be to cast the shape desired in flat or bivalve molds and then hammer and file the edges. Another method is to use flat sheets of metal, cut out the desired shape with metal snips, metal saws, or groove-and-snap techniques, and then file the edges to sharpen and shape them. The choice of method used depends on the ability to produce the high temperatures necessary to melt and cast metal, the availability of sheet metal, the existence of tools like saws and snips, and the prior and existing traditions of how metal is worked to make these and other objects. The traditions of production process or the *technological style* of process, can be just as much a factor in choice of method as technical knowledge or availability of materials and tools, as discussed in Chapter 5.

Knowledge of the production process also allows us to identify requirements and rhythms that affect both the organization of production and the relations between production and consumption, including distribution and demand. For example, stone tool production by North American flintknappers often involved an initial stage of rough shaping at the quarry site, to allow more efficient transport of a smaller stone "blank" to sites where the final stone tools were produced as needed. These tools themselves could be further modified to produce new tools as required (Bradley 1989). Knowledge of the possibility of separating the stages of production, and transporting the interim products to other locations, thus helps in our attempts to identify ancient production, trade, and consumption patterns, as we can expect to have some stages of production missing at some sites. The widespread trade in metal ingots around the Mediterranean during the Bronze Age (Tylecote 1987; Henderson 2000) is another example of the importance of identifying *segregation of production*—the locational separation of stages of production—in order to see patterns of trade. A third well-known example is the effective use of segregation of production stages by Chinese porcelain makers, with different production stages in separate locations employing separate workers, as a means of controlling production knowledge and maintaining their monopoly over the production of porcelain. Intensive efforts by Europeans to re-create or steal these techniques were greatly impeded for some time by this practice of production segregation, in a fascinating example of early industrial competition, directed experimentation, and material development (Kingery 1986).

The focus in the next two chapters is thus on outlining production processes to provide an introductory background. However, besides the techniques involved in production (how production occurs), researchers also examine the location of production stages, distribution and use spots, and discard areas (where); the people involved in both production and consumption (who); the types of objects produced (what); the timing of production, distribution, use, and discard (when); and the choices made in production and consumption (why). Therefore, I have selected brief examples of these other aspects of production and consumption for each category. Given the very limited space available, I have made an effort to discuss a different aspect of production organization or consumption in each case in Chapter 3. For example, although social aspects of exchange and distribution have been studied for every category of craft, I only discuss it for stone. Instead of exchange, I use the space in the fibers section to provide a discussion of scheduling conflicts for both workers' time and resource use. Aspects of production organization and consumption for the crafts in Chapter 4 are discussed in several of the examples in Chapters 5 and 6, which provide more extensive thematic studies from the perspective of technological systems as a whole.

CLASSIFICATION OF CRAFTS

While my primary classification scheme for Chapters 3 and 4 focuses on the principal material used to produce objects (stone, fiber, wood, metal, clay, glass), I secondarily use the type of production process to group these classes into extractive-reductive crafts (Chapter 3) and transformative crafts (Chapter 4). This grouping follows work by Italian and North American archaeologists (Pracchia, et al. 1985; and subsequent modifications by several researchers outlined in H. M.-L. Miller 2006). It is also very similar to

Moorey's (1994) division of Mesopotamian crafts types into those employing only mechanical modification of the raw materials (my extractive-reductive crafts) and those employing chemical and structural alterations to the raw material (my transformative crafts). I focus on differences in the production processes of these crafts because I am most interested in technology. I could instead have used a typology focused on the nature of the material, such as organic versus inorganic materials, as is done by many archaeologists and by conservators who are primarily concerned with the different preservation characteristics of organic and inorganic materials.

As noted previously, archaeologists tend to classify technologies on the basis of the principal material employed, such as metals, stone, clay, or bone, rather than on the basis of function, such as containers, cutting tools, or ornaments. This is because it is usually relatively simple to sort objects visually by general material type, but much more difficult to identify the function of an unknown object. Similarly, it is relatively rare to use the production process employed in manufacture as a way to initially classify objects, but once objects are divided into materially-based groups, it is common to use production processes to further classify them (e.g., Chazan 1997: 733 for knapped stone; Adams 2002: 11–16 for ground stone; Seiler-Baldinger 1994 for textiles).

This chapter deals with extractive-reductive crafts. *Extractive-reductive* crafts use extractive or reductive processes such as chipping, grinding, carving, and twisting to process raw materials into finished materials or objects. These crafts frequently also employ methods of joining such as twining, weaving, pegging, and gluing to build composite objects like clothing or furniture. This category includes stone production and woodworking of all types, from the production of tools to ornaments to buildings. Extractive-reductive crafts also include shell, bone, antler, leather, fur, bark and feather working, as well as basketry and textiles of all sorts. Note that if I were using functional rather than material-based craft divisions, architectural and building crafts would be included in this category, although the materials used can be either extractive-reductive (stone, wood, reeds, unbaked clay) or transformative (baked brick, plaster, concrete). Chapter 3 covers three of the main material-based groups of extractive-reductive crafts: stone, both chipped and ground; fiber, including both basketry and textiles; and wood and other sculpted organics such as bone, ivory, and shell.

Transformative crafts are discussed in the next chapter. *Transformative* crafts transform raw materials through pyrotechnology or chemical processes to create a new material. These new materials have had their basic physical or chemical structure transformed by human action. The vast majority of this type of ancient craft involved the application of heat – pyrotechnology. Pottery production is the most ubiquitous of these crafts archaeologically; related clay-based crafts include baked brick manufacture and the creation of

Extractive-Reductive Crafts

baked clay figurines, clay pipes, and other objects. This category also includes metal production of all types, as well as silica-based crafts from faience to glass to porcelain. Although not discussed in this book, the production of plasters is another case of transformative pyrotechnology. The dyeing of stone or cloth are transformative stages of production, if the actual material is chemically transformed. A similar example is the chemical etching of stone, metal, and glass. Chapter 4 covers three of the main material-based groupings of transformative and related crafts: clay-based crafts, such as pottery and brick production; metals, primarily copper-based and iron-based types; and glazes and glass, including faience.

Would that the world of technology were actually so tidy. Of course, few objects are made from only one type of material. The reality is that many objects are the composite products of what archaeologists would classify as multiple crafts. Even for objects composed of a single material, however, many crafts have both extractive-reductive and transformative stages. These categories are actually two ends of a gradient, from crafts for which almost all production stages are extractive-reductive, like basketry or shell working, to crafts for which the definitive stages are transformative, like the production of cast metal objects or the production and etching of glass items. Between are *bridging* crafts, those with substantial use of both extractive-reductive and transformative stages. A common bridging craft in ancient Eurasia is the glazing of stone, as described in the discussion of the creation of new materials in Chapter 6. Beads and small objects made by extractive-reductive processes from talc and other stones were covered with a colored glaze created by (transformed by) the application of heat. Other types of stone object production might also be classed as bridging crafts, as they involve significant heat treatment of the raw stone; the production of red carnelian from certain agates is such a case. Dyed textiles would comprise a bridging class, where the fibers and cloth were produced by extractive-reductive methods, and the colors created by chemical treatments that transformed the properties of the fiber. Transformation of color seems to be a common reason for adding a transformative process to what is otherwise a largely extractive-reductive craft, although by no means the only reason.

In addition, the classification of a craft into extractive-reductive or transformative may vary if it is practiced in different manners in different societies. For example, metal production is predominantly transformative in metal working traditions that primarily employ casting or joining with molten metal, as was the case for most of the ancient world. However, some metal working traditions, notably that of pre-European North America, were entirely extractive-reductive, employed methods of metal fabrication similar to stone working crafts, as is discussed in Chapter 5 in the section on Style and Technology. Such differences in *production style* are exactly why a division of crafts into extractive-reductive and transformative types is useful, if not precise, because it draws attention to significant differences between traditions in the practice of entire crafts. By using a production process-based classification system rather than simply a material-based or preservation-based classification system, this very different treatment of the same material, metal, is forced on my notice, and I am able to immediately recognize these variations in technological style. This example shows the tremendous amount of choice available in the practice of technology, even within the considerable constraints imposed by the physical aspects of metal production. Nevertheless, the choices involved in achieving a particular type of end product are usually considerably limited by physical constraints, which is fortunate for the researchers trying to reason backward from object to production process. Such physical constraints are most frequently discussed for the study of stone or lithics.

STONE/LITHICS

It is appropriate to begin with stone, as stone artifacts define the beginnings of archaeological study; the first stone tools are the earliest remnants of cultural behavior that have been preserved. Traditionally, stone artifacts are divided into *flaked or knapped stone*, produced by hitting the stone with another object in a process variously called flaking, knapping, or chipping, and *ground stone*, produced with some knapping but primarily by pecking (pulverizing) and grinding. However, other sorts of production techniques are also employed in the manufacture of stone objects, such as grooving and snapping, or cutting. As will become apparent in this section, there is considerable overlap of techniques in the production of "knapped" and "ground" or "cut" stone objects. Therefore, I prefer to discuss stone working as a whole, employing a modified version of Hodges' (1989 [1976]: Chapter 7) terminology for the various types of shaping techniques: *knapping, cutting, pulverizing,* and *abrading*.

Lithics is a term is used to refer to both the end-products (tools, objects) and the by-products (debitage, debris) of stone object production. In the general literature, the term lithics is most frequently applied to knapped stone objects and their production debris, and less often used to refer to ground stone or cut stone objects. Odell (2004), Inizan *et al.* (1992), Whittaker (1994), and Andrefsky (1998) provide different introductory overviews of the process of lithic production and analysis. Inizan *et al.* (1992) also includes a multilingual vocabulary of lithic terms, while Odell includes methods of tool functional analysis, both use-wear and residue analysis. These volumes are primarily focused on knapped stone production, except for short sections in Odell. Adams (2002) is thus a very welcome addition for ground stone tools, particularly for the analysis of object use and for North American assemblages; several more specific studies of ground stone production and use are cited below. Hodges (1989 [1976]) has a brief but informative outline of techniques used for sculpture and masonry. The journal *Lithic Technology* is a good place to search for specialist-oriented analyses, but almost every archaeological journal regularly publishes articles related to lithic studies. It is also well worth investigating the many excellent websites devoted to lithic studies. An enormous number of edited volumes discuss themes of lithic analysis and interpretation; for example, the recent volume edited by Kardulias and Yerkes (2003) contains a variety of case studies on topics from sourcing to economic control from around the world. Lithics are also the main type of artifact studied with the *chaîne opératoire* approach described in Chapter 2.

The general production of stone objects employs the following stages (Figure 3.1):

1. Collection of stone (selection, quarrying)
2. Preliminary processing of stone (cortex removal, possible production of blanks), including an additional transformative heating or dyeing stage if necessary, to improve working qualities and/or change the color (if present, this stage can also occur earlier or later in the sequence)
3. Shaping of stone objects, employing (a) knapping/flaking, (b) cutting (sawing or groove-and-snap techniques), (c) pulverizing (primarily pecking), and/or (d) abrasion
4. Finishing stages, which typically employ abrading techniques (polishing, smoothing), but can also include knapping for sharpening edges, or transformative techniques such as dyeing.

COLLECTION AND PRELIMINARY PROCESSING

From very early times, humans (and closely-related extinct species) have desired specific types of stone, collected them from their source locations, and distributed them across the landscape by a variety of mechanisms. For example, stones were traded between people living at the source and others living farther away, carried by traders, herders, or pilgrims traveling between the two locations. Stones were also collected by hunter-gatherer or herding people at a source that was one part of their annual seasonal round. So the finding of an "exotic" stone at an archaeological site is an immediate clue that some sort of connection existed between the find location and the source location; the task of the archaeologist is then to determine the nature of that connection. The location of the stone source is thus of great interest to archaeologists, as an initial clue to the possible extent of regional connections.

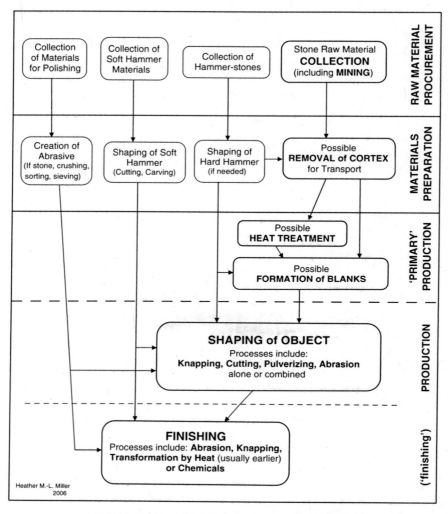

PRODUCTION PROCESS DIAGRAM FOR STONE

FIGURE 3.1 Generalized production process diagram for stone (greatly simplified).

Researchers have used both macroscopic and chemical analytical techniques to *source* or *provenience* stone, to determine the location from which raw materials were collected, especially for flints/cherts and obsidians (e.g., Ericson and Purdy 1984; Luedtke 1992; Shackley 1997, and numerous articles in *Archaeometry* and *Journal of Archaeological Science*).

Functional (physical use) characteristics affect the choice of stone used to make an object, depending on the requirements of the end-product. The different mineral characteristics of different types of stone, such as mineral composition, grain size and structure, affect their texture, hardness and toughness (resiliency or durability). For different sorts of stone objects, different functional requirements will be desired. Cutting or piercing tools requiring a sharp edge are preferentially made out of stones whose mineral structure allows easy formation of sharp edges, such as obsidian, chert and chalcedony (agate or jasper). Less brittle (tougher) stones like quartzite can also be used for cutting, but it is more difficult to flake these stones due to their more blocky mineral structure, so if formation of a sharp edge is the main criteria, these stone types will only be used if the easier-to-flake stone types are not available. Other tools for which toughness is the primary criteria, such as pounding implements, hoes, adzes, and often axes, will be formed from stones which do not flake easily on impact, such as basalt, granite, or jade. Grain size (coarseness or texture) can be as important a factor as toughness in choosing materials to make effective grinding implements (Adams 2002). For small sculptures or vessels, stones that do not shatter when shaped yet which are relatively soft and fine-grained are often preferred, such as soapstones and related minerals.

The choice of stone type is affected not only by the intended physical function of the end product, however, but also by desired colors, patterns, and reflective characteristics, and by any cultural symbolism associated with particular stone types. This is obviously the case for stone used to make jewelry, decorative items, or religious objects, but it may also be important for tools in some cases. For example, a hunter might consider a projectile point made from petrified wood to be a far more effective weapon for killing particular animals than any other type of stone, due to its special symbolic powers. He would thus go to great lengths to acquire this stone, even though its flaking and penetrating properties might not be any more physically efficient than more easily available lithic materials. Similarly, minimum standards of strength and workability are necessary for architectural or sculptural stone, but beyond those requirements particular types are often chosen with specific cultural associations in mind, especially for public buildings.

The collection of stone can be as simple as picking up cobbles in a stream bed, a physical task requiring few tools but one that can entail considerable expertise to choose the correct materials. However, in many cases, preferred types of stone had to be *quarried*; that is, dug, chipped or broken out of sedimentary or block deposits. Cobbles deposited by water or glaciers are found buried in sediments from which they can be collected by digging with digging sticks if deposits are shallow, or more elaborate tools if deposits are deep. Stone is also found in block deposits from which smaller portions must be extracted, again using a variety of tools. The degree to which fire-setting was used to quarry stone in the past is much debated; it certainly did occur, but the prevalence and extent of its use is not clear. However, for quarrying large blocks of architectural or sculptural stone, more precise methods of splitting than fire-setting would have been used. Hodges (1989 [1976]: 108–111) discusses the tools and processes employed in splitting and working stone used for architecture and sculpture. Stone quarrying could also involve extraction techniques similar to those used for the quarrying of metal ore, as discussed in Chapter 4.

Frequently, the stone collected by all of these methods was reduced to smaller pieces at or near the collection site. The weathered outer skin or *cortex* of the stone as well as any flaws or unwanted mineral deposits would be removed to decrease weight for transport. Therefore, collection sites often contain large quantities of *primary flakes,* flakes with cortex covering a large proportion of their surface, as well as generalized blocky and jagged waste material. *Blanks* might even be produced near the collection site, a block of stone of the approximate size and shape needed for producing the planned stone object or architectural piece. Bradley (1989) beautifully illustrates the production of blanks for early North American stone tools, as well as the potential versatility and transport advantages of blanks. Biagi and Cremaschi (1991) provide an example of industrial-scale blank production of chert cores for distribution to towns and cities across the Indus Valley, where the final blades and drills were produced.

However, as archaeologists know all too well, people had to balance various requirements in making choices, so that there were times when it was a disadvantage to remove the cortex of stones at the collection site, even though this greatly increased transport costs. For example, archaeologists conducting ethnographic and experimental studies of stone bead production in the town of Khambhat (Cambay) in India, found that entire agate nodules were typically brought back to Khambhat from the source area (Kenoyer, *et al.* 1991, 1994; Roux 2000). Only a small window would be chipped in a nodule at the quarry site, to determine its likely quality (Figure 3.2, left side). This was done for a number of reasons. Most of the nodules were carefully heat-treated, to improve working qualities and change the color of the agate, and this was done by each workshop group prior to any knapping of the stone. Presumably each craftsman preferred to oversee this stage of the process himself. In addition, it was said that the practice of doing all work in the town, and transporting the agate in nodule form from the mines, developed because of the long history of banditry in the region (Kenoyer, *et al.* 1994). Bandits preying on travelers and traders were less likely to steal bags of raw nodules than valuable stone beads or even semi-finished blanks.

FIGURE 3.2 Products and some debris from stages of production for agate/carnelian beads; progression of production stages is from left to right.

This is a relatively recent case, where we have ethnographic accounts to alert us to these unusual choices made in the location of this preparatory stage of production. We seldom have this sort of information for the past, but such information provides essential information about the nature of economic and possibly political situations affecting the production of these objects. This does not mean that the same thing *did* happen in the past. But it does mean that archaeologists need to consider such possibilities by checking the archaeological data for additional information. In this case, archaeologists need to check both collection sites as well as other settlement-based production sites for the presence or absence of primary flakes and other sorts of *debitage* (debris from stone working), before they assume that cortex removal primarily took place near the collection site. This is why it is so important that source areas for raw materials are well-documented by archaeologists, and why avocational flintknappers are encouraged to remove their own debitage and generally avoid disturbing old quarry sites (Whittaker 1994).

In the Khambhat example, the heating of agate nodules is an important stage of production, done prior to any shaping of the stone, or even the removal of cortex (Figure 3.3). For some types of stone and some purposes, ancient craftspeople used heat or chemicals to transform the stone, changing its working properties and/or its color. The most widely known transformative technique for stone is the heat treatment of chert or flint, thought to have been used to decrease the number of tiny flaws in the material and thus improve its

FIGURE 3.3 Heat treatment of agate nodules inside pottery vessels, using rice husks as fuel in a reducing atmosphere. *Photo courtesy of Lisa Ferin.*

working properties (Grifiths, *et al.* 1987; Luedtke 1992; Purdy 1982). In these circumstances, heat treatment would obviously take place prior to the shaping stage (Figures 3.4 and 3.1). As in the Khambhat example, color change or enhancement may have been an additional or even the primary goal of heat treatment. Heat treatment would still be done prior to working of the stone, to improve the working quality of the stone as well as change the color. Heat treatment might also be done again after the object was finished, to deepen the color (Matarasso and Roux 2000). Color change or design enhancement was the goal of most chemical methods of transformation, such as the

(a)

FIGURE 3.4 Experimental heat treatment of chert fragments: (a) placed on sand.

Extractive-Reductive Crafts

(b)

(c)

FIGURE 3.4 (*Continued*) Experimental heat treatment of chert fragments: (b) covered in more sand; (c) buried under dirt and a fire set.

chemical dyeing of semiprecious stones like lapis, or the chemical etching of designs on stone. However, these chemical transformative stages typically took place after the shaping stage, often as one of the last stages of production, as they were focused on the appearance of the final surface (Figure 3.1, bottom).

SHAPING AND FINISHING METHODS

The shaping methods employed depend largely on the type of stone being worked, particularly the hardness and the toughness (resilience) of the stone. Obsidian and chert are very hard, but are very brittle, not tough or resilient. Therefore, they are best shaped by the rapid process of knapping; other shaping techniques are much slower and can cause them to shatter. Basalts and granites are also hard but are relatively resilient, so that while they can be shaped with difficulty by knapping, they can also be formed by abrading or by pulverizing techniques such as pecking. Jades are similarly very hard and very tough, and so are best formed by pulverizing and abrading. Soapstones are soft and resilient, not brittle, so they are easily formed by cutting, but do not knap well.

Once the type of stone is chosen, this largely dictates the methods of shaping and finishing the material: knapping, cutting, pulverizing, and/or abrading. Lithics are noteworthy for the degree to which physical constraints imposed by the properties of the stone restrict choices in production methods. Lithic specialists are especially interested in the physical constraints restricting choice, as the restriction of possible production processes allows researchers to identify production techniques, sequences, and even end-product shapes on the basis of the debris left behind (Kardulias and Yerkes 2003). Innumerable experimental and ethnographic projects have been dedicated to this end, including the ethnographic and experimental work done in Khambhat (Cambay), India, described above.

Knapping

Knapped (or flaked) lithics are shaped by hitting the stone (percussion) and by applying pressure. To briefly introduce the extensive terminology for knapped lithics, the term *core* refers to the main block of stone, from which *flakes* are struck off. Percussion techniques include hitting the core directly with a hard object (stone, metal) or a soft object (wood, bone, antler) to detach flakes (Figures 3.5a and 3.6). These are called *hard hammer direct percussion* and *soft hammer direct percussion*, respectively. *Indirect percussion* occurs when an object is used to strike a wooden or bone punch placed against

Extractive-Reductive Crafts

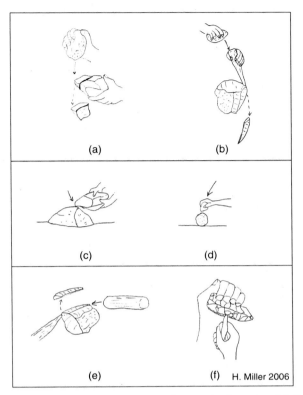

FIGURE 3.5 Illustrations of various percussive techniques for knapping: (a) direct percussion with hard hammer; (b) indirect percussion; (c) block and anvil technique; (d) bipolar technique; (e) inverse indirect percussion; (f) pressure flaking. (Redrawn after Kenoyer n.d. and Odell 2004.)

the core, detaching flakes from the core (Figure 3.5b). Although typically thought of as a Paleolithic method, the use of hammers striking metal spikes to quarry building stone or mine minerals is also a form of indirect percussion. Even more complex systems exist, such as *inverse indirect percussion* (Figures 3.5e and 3.7a & b) (Kenoyer, et al. 1991, 1994; Pelegrin 2000; short videos of this technique are shown on the CD in Roux 2000). *Anvil* or *block and anvil* techniques are also percussive, but in this case the core is hit against a stationary stone called the anvil (Figure 3.5c). A similar process is employed in the *bipolar technique*, used for small pebbles, where the pebble is placed on an anvil (sometimes held with the fingers), and hit with a hammerstone (Figure 3.5d). This can create orange-segment-shaped primary flakes, but the technique is especially characterized by the presence of two points of percussion, one on each end of the flake; hence "bipolar" (Figure 3.8).

FIGURE 3.6 Creation of building stone blocks by knapping, using direct percussion with iron hammers.

FIGURE 3.7 Creation of agate beads by knapping using inverse indirect percussion: (a) Inayat Hussain, mastercraftsman, and (b) workman in the Keseri Singh workshop. *Photos courtesy of Lisa Ferin.*

This technique is typically used in areas where knappable stone is scarce, as seen in the use of small river pebbles of obsidian in the areas around the Columbia River of the Northwest Coast of North America (Liz Sobel, personal communication). For additional terminology, Odell (2004), Inizan *et al.* (1992), Whittaker (1994), and Andrefsky (1998) contain extensive details

Extractive-Reductive Crafts

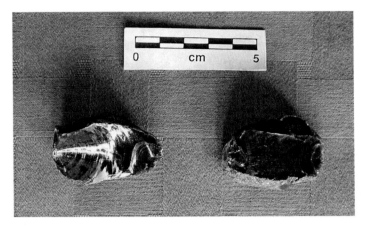

FIGURE 3.8 Flake of obsidian created by bipolar technique; note the two points of percussion.

of various production processes for knapped lithics, as well as illustrations of tool and debris types.

These percussive techniques are used to shape stone objects. Pressure techniques may then be used to refine the shape, sharpen edges, or smooth surfaces (Figure 3.5f). In *pressure flaking*, a small tool (usually an antler, bone, metal, or wooden rounded point) is laid against the knapped edge and pushed gently to remove a small flake—the flake comes off of the underside of the knapped edge. By varying the angle at which the tool is held, the flintknapper can take off flakes of varying thickness and length. A skilled flintknapper can remove very long thin flakes which extend a great distance across the knapped object, as is seen in projectile points and knives of the Paleoindian and Upper Paleolithic periods in North America, Eurasia, and Africa. Pressure flaking is generally used to reduce the thickness and refine the shape of such points and knives, to make a serrated edge, or to sharpen a dull edge. Many of these effects can also be achieved with percussive techniques as well, so care must be taken when analyzing knapped artifacts. Knapped stone objects can also be smoothed by grinding techniques after they are shaped; many agate and jasper beads, bowls, and other objects produced in South Asia today are still shaped by knapping, although no traces remain on their final highly polished surfaces (e.g., Figure 3.2, right side).

Cutting (Sawing, Drilling, Groove-and-Snapping)

Some types of stone will not flake well, due to their mineral structure. The softer varieties of such stones, including soapstone (talc or steatite), alabaster, turquoise, and lapis lazuli, were and still are shaped primarily by cutting methods (Figure 3.9). *Cutting* is the shaping of a stone using a tool that is much

FIGURE 3.9 Stages in production of soapstone (talc/steatite) beads (right to left): pieces cut with a copper saw, drilled, strung, and edges abraded as a unit on a wet grinding stone.

harder than the stone itself, so that the tool can cut the stone without use of an abrasive, and without much wear on the tool itself (Hodges 1989 [1976]: 107). The tools used could be made of harder stones, including flint or chert, or of metals. Relatively soft architectural stones, such as sandstones, limestones and slates, have also frequently been processed using cutting techniques, particularly once metal saws and chisels were available. Drilling of soft stones (and organic materials) was often a form of cutting, especially once metal drills were available. However, harder varieties of stone were primarily *abraded* in the past, as discussed below.

Groove-and-snap is a particular type of cutting technique very commonly used around the world in many time periods, especially when working with stones that had a blocky, laminar structure. A groove would be cut into the stone, usually by repeated incising with a chert flake or metal point. Once it reached sufficient depth, the stone would be snapped along the groove. A corresponding groove might be cut into the opposite side, to ensure a proper break. This technique was frequently used to make disc or cylindrical beads of soft stone, shell, and other organic materials.

Pulverizing (Pecking)

Objects made from tough stones can be further refined in shape by pulverizing or pecking, hitting the stone with a hammerstone at approximately a 90° angle to create a multitude of small pits in the stone, to slowly shape and even out the surface. This technique is most commonly described for the manufacture of "ground stone" objects, such as grinding stones and adzes.

Extractive-Reductive Crafts

These are frequently roughly shaped by flaking, using the percussive techniques described for knapped stone (Figure 3.5). Quite often, the tough nature of stones selected for the production of ground stone objects requires the use of hard hammer direct percussion or block and anvil techniques, removing large flakes. Another approach, used whenever possible, is to select a natural stone cobble or a stone block of approximately the desired shape and size, as described in the vignette from Papua New Guinea below. However, pecking was also used to shape or decorate other sorts of objects made from tough stones, such as architectural and sculptural pieces. Such objects might be roughly formed first by cutting, or by indirect percussion with other stone, wood, or metal tools.

Abrading (Grinding, Smoothing, Polishing, Drilling)

Finally, the shaped stone can be abraded or "ground" to create a smooth, even surface. Abrading can employ the rubbing of the stone object against another stone to wear down the surface, or another material (stone, terracotta, wood, leather) can be rubbed against the stone object to abrade it (Figure 3.9). Quite often abrading involves the use of small hard particles such as sand or corundum as an additional abrasive. These particles can be applied alone to the surface of the stone object, but are typically part of a water- or oil-based paste or slurry, or are even imbedded in the material used to rub the object (as with modern sandpaper). Progressively finer abrading materials can be employed, to create an increasingly smooth, polished surface as desired.

Drilling is also often a type of abrasion rather than cutting, particularly the drilling of stone objects. Variations on simple hand or bow drills have been used worldwide for drilling stone, wood, shell, bone, and other materials (Figure 3.10). Given the disparity in hardness, the drilling of wood or soft stone with a metal drill involves cutting. However, drilling hard stone with stone, copper, or even iron often involves abrading, given the much reduced disparity in hardness. "Sawing" can also involve abrasion rather than cutting, if string or wire or a metal disc is used with an abrasive (e.g., Foreman 1978). Abrading was the most common method of "cutting" hard stones in the ancient past; cutting usually was possible only with relatively soft stones, until the development of steel or diamond-impregnated cutting implements.

Production Stages

It should be clear from even this brief discussion that the techniques used in processing stone are complex and varied, and the order in which techniques

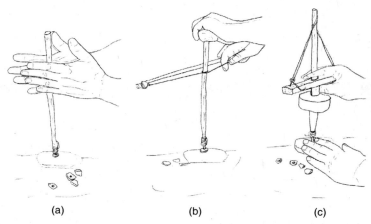

FIGURE 3.10 (a) Hand drill, (b) bow drill, and (c) pump drill. (Drawn after photographs in Foreman 1978.)

are used can be quite variable. For example, a stone bowl might be made by rough percussive knapping followed by pecking and abrading. Alternatively, the rough shape might be cut with a saw, the center cut out as a plug using a tubular drill, the surface abraded, and the bowl dyed or etched with decorative patterns. Even when basically the same techniques are used in the production process, there is variability possible in the *order* of the stages, even at the same site, as Vidale and associates have shown for stone bead production (Vidale, et al. 1992; Vanzetti and Vidale 1994).

Sometimes the end-product of the processes described above would be a finished object, a grinding stone or a statue. More often, the stone objects produced were combined with other materials to create a finished object. A stone adze would be hafted to a wooden handle using fibers and possibly adhesives (Figure 3.11). As Bleed (2001: 155) discusses, this process of hafting is a "surprisingly complex undertaking" and a subject of technological investigation in its own right. Similarly, stone beads would be strung with other materials to create jewelry; colored stone shapes might be inlaid into wooden furniture; and stone blocks could be combined with wood, brick, metal and plaster to form buildings. The vast variety in production process organization is apparent; the result of this variety is both bane and boon for archaeologists and other technology specialists. On the one hand, understanding the web of processes and relationships involved in producing even a simple object can be an overwhelming task. On the other hand, this very complexity is the window we have into social, political, and economic relationships, as well as cultural traditions of the way things 'ought' to be made, distributed, and used.

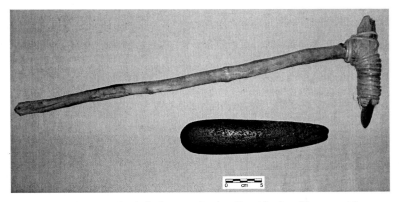

FIGURE 3.11 Small stone adze hafted to wooden handle with plant fibers, used by women for weeding gardens, and unhafted large stone adze, used by men for chopping wood both from central New Guinea.

ORGANIZATION OF PRODUCTION; CONSUMPTION

The *organization of production* refers to the interconnections between people and objects during the process of production, as is more fully defined in Chapter 1. "People" includes craftspeople and managers or owners (if different from the actual producers), and "objects" includes both the objects produced and the tools used to produce them. Archaeological approaches to the organization of production, including labor issues, are described in more detail in Chapter 5. Here I will instead focus on aspects of consumption, with an example showing how technological systems include not only production and organization of production, but also distribution and use. I am broadly defining *consumption* to include the distribution of finished objects, as well as their use, re-use, and discard.

The processes of distribution are much studied for stone, in association with the long history of interest in the proveniencing (sourcing) of stone materials. Once an end-product is finished, or more likely throughout the production process, decisions are made about distribution to consumers. Such considerations are expected for large-scale, mass-production systems, where products are often standardized and distributed on a large scale. The distribution of obsidian tools and objects throughout Mesoamerica provides numerous examples of such systems through various time periods under a number of political systems (Cowgill 2003). Such was also the case for chert blade production in the Harappan Phase (Mature Harappan) period of the Indus civilization, where throughout the region there is a shift to nearly exclusive use of chert blades made from Rohri Hills bullet cores, cores which were produced at the source

FIGURE 3.12 Roger Lohmann (left) and Salowa Hetalele (right).

location and distributed throughout the Indus region (Biagi and Cremaschi 1991). However, large-scale mass-produced systems are by no means the only case where producers consider distribution during the production process, as the New Guinea example below illustrates.

As discussed in Chapter 2, ethnographic accounts form an essential source of archaeological information about production and consumption processes. The account reported here briefly outlines one example of stone adze manufacturing and distribution in a small-scale society in central New Guinea (Figures 3.11 and 3.12). In this case, the question "How did people make stone adzes?" elicited a response far beyond a simple description of the production process. Salowa Hetalele's answer shows how producers considered the potential value of objects, as well as the social networks of access and distribution, as a part of stone adze technology.

Ground Stone Adze Production in central New Guinea
**Sun. 12 March, 1994. Yakob Village, Duranmin,
Sandaun Province, Papua New Guinea**

The speaker, Salowa Hetalele was born about 1947, and was a resident of Kalu River before Australian contact in 1963. His father was Asabano (Duranmin) and his mother was Akiapmin (Towale), two related ethno-linguistic groups in central New Guinea. This excerpt was recorded and translated from the Tok Pisin by the interviewer/ethnographer, Roger Ivar Lohmann.

Stone adzes. They are big, at Tahu River at the head of the Wario River, they made them from a rock; they prepared a big rock. It's a huge rock, big as a house. Now they made a fire. Break firewood, bring it, and when it was ready, okay, they made a platform now. The wood they cut and brought, planted it, back and forth, back and forth, making a platform on top. They got soil and put a layer on top of the platform, just like a hearth. Then they made a fire on top [of the platform]. The fire burned and burned, and it was like a cloud exploding [i.e., thunder]. This done, they said, "Alright, the stone adzes are broken off now."

Many men gathered and made the fire. Time passed now; they made a fire, and "break break break, break break." It was like that. The fire rose and the rock broke. Now it kept breaking as the fire burned, and eventually the fire went out. Now, one stone, someone took it. Took it on top now, and he hit it now. He beat it until it was like a banana—soft. "Break, break, hit," he hit it, and the others stood far back. If you stood close by you'd be finished! Having hit the stones for some time now, all the stones broke and fell down. They broke and fell, covering the ground in a scatter.

Now they gathered and went to get these pieces of stone, get them, standing in a line, advancing like that. Then they hit them. Hit them and broke, broke, broke them on and on, one, two, three, four, five, six, seven, eight, nine, ten, it went up to like twenty or a hundred. They were hitting it now. Hitting continuously, and when it was the right shape, they got a rope now and fastened it. Some took net bags now and put it inside. They carried it to the house and left it there; it was already dusk.

Tomorrow morning, getting up, they would say, "Brother, look at mine, it's good, and mine is very long. Look at it. And yours is short! Mine is better." They'd chat like that. "And you wait and see, mine will later be enough to pay for a pig." And one would say, "Mine will pay for a bow." Later he would say, "Mine will later pay for a dog." They'd sit and talk, hitting them now. "Kana, kana, kana, kana [the sound of hitting]." Hitting and hitting, and turning and hitting on and on, until they were finished.

Okay, they went to the river now, and got a stone and rubbed them back and forth. They polished and polished, and eventually, look—it was really sharp! Alright, they cut wood and branches and brought them, put them aside, got vine rope, tied it now, okay, and now they could use the adze to cut trees.

This ended with me. People of my age, along with Omahu, Bledalo, and Yalowad; they saw it, and my father Madfe; he saw it. It ended with us. We saw it. And Omahu and Bledalo, those two used stone adzes. I think they used them a little. Not me, I just saw them.

And making them, that was at Wario, the Tahu River people made them. The people at Kalu River couldn't make them. The people at Kienu River couldn't make them. Just at Wario, at the head of the Tahu River; there they would make them. At the head of the Nena, the Mei Rivers, they made them. They made them and sold them to Lembana, and Mondubanmin, and Oksapmin, and here. Telefolmin, we don't know. Did they, too sit and make stone adzes? I don't know. And Oksapmin, we sent them there, and to Sisimin and Sugamin and Duranmin. It's the Akiapmin and Towale, they sent them up here. They were the source of all the stone adzes. Duranmin [Asabano] people are related, so they would go and watch, or go and make them, and carry them back up [the mountain to their home]. But Kienu River people and the Sugamin, they couldn't make them.

For Salowa, as for scholars interested in technology studies, groundstone adze production extended beyond physical techniques of manufacturing to include who could make stone adzes and what they could be traded for. In cases of larger-scale production of lithic objects for trade, production was often very explicitly linked to demand. Ethnoarchaeological studies of chipped stone bead production in Khambhat, India, found differences in the way different scales of production reacted to demand (Kenoyer, et al. 1991, 1994; Vidale, et al. 1992). For the bead-makers who were large-scale merchants, massive quantities of raw materials (agate nodules) were purchased at once, sorted and dried, and might be heated or not prior to storage until needed for production. In contrast, bead-makers who were small-scale entrepreneurs purchased raw materials in smaller quantities and hired additional workers as needed for particular jobs. There were also frequent instances of merchants commissioning special orders. Kenoyer, Vidale, and Bahn linked these and other ethnoarchaeological patterns to archaeologically-recognizable signatures, and used them to interpret patterns from archaeological sites of the Indus civilization (also see Barthélémy de Saizieu 2000).

In Southeast Asia, the production and consumption systems for agate beads during the first millennium BCE have been strongly linked to arguments about the mechanism for demand, as a factor in emerging hierarchical social and political systems (Theunissen, et al. 2000; Bellina 2003). One viewpoint models Southeast Asian state formation as heavily influenced by contact with Indian states, particularly via trading networks. In this model, demand for exotic trade goods, including agate beads from South Asia, was a status-raising technique employed by local Southeast Asian elites, who controlled the trading contacts. A more recent model draws on proveniencing of some Southeast Asian agate beads to Southeast Asian as well as Indian deposits, indicating local production as well as other origins (Theunissen, et al. 2000). This evidence is used to support models of more complex systems of production and exchange for status-marking (prestige) items, with Southeast Asian demand also influencing the Indian producers (Bellina 2003). In this case, indigenous elite class formation occurs with Indian influence as an overlay, manipulated by local elite for their own ends, rather than a more passively adopted impetus.

After distribution, consumers use, re-use, and discard objects in a multitude of ways, all of which are part of the technological system relating to the object. Use, re-use and discard are often studied for stone tools, such as in analyses of the breakage patterns of different kinds of points of stone, bone, and antler (Knecht 1997). Researchers in many parts of the world also have studied the modification of blanks and used or broken tools to create a series of new tools. Bradley (1989) illustrates this process for a hypothetical hunting trip in ancient western North America, showing the creation of a sequence of needed tools, including a projectile point, which is broken during the kill, a butchery knife, a hide scraper, and a replacement point. Rolland and Dibble (1990; Dibble 1995) and others have examined the modification of heavily used or broken Mousterian tools into other tool types to understand both resource exploitation and the relationship between different "types" of stone tools, as part of the continuing great debates of the European Paleolithic initiated by the famous differences of interpretation between Bordes (1961; 1973) and the Binfords (1966).

Stone tools are themselves used in production of different types of objects, such as the use of a stone adze to produce various wooden objects. Examination of the stone adze can tell archaeologists something about how the adze was made, from traces of production processes such as knapping and abrading. In addition, however, examination of the stone adze for traces of use can also provide information on how the adze was used, including working wood, scraping leather, butchering animals, and tilling fields. Usewear analysis examines both macro and microwear traces, using a range of microscope types; even more information is available in many cases from residue analysis for chemical and physical traces (Odell 2004). Such usewear and residue studies are particularly powerful when used together, and are important in understanding the consumption part of the technological system for the stone tools themselves. Usewear and residue studies are also extremely important when trying to recreate the production processes associated with the working of poorly preserved materials, such as wood, leather, and fiber.

FIBERS: CORDAGE, BASKETRY, TEXTILES

Although stone artifacts define the beginnings of archaeological study, as the earliest artifacts preserved, most archaeologists suspect that either fiber or wooden artifacts were truly the earliest modified materials, but are simply not preserved. The earliest fibers could have been used for tying or carrying bundles, or for stringing shells or seeds to make the first ornaments, since there is no reason to believe that tools are earlier than ornaments, as discussed in the beginning of Chapter 6.

Basketry and textiles are generally divided into two different craft groups, but from a technological perspective, there is a great deal of overlap. Both basketry and textiles begin with production of fiber strands made from a wide range of materials, but most often from plant fibers. These fiber strands are then twined, twisted, or woven together to form a "fabric" of some kind, where *fabric* is defined simply as material made from fibers. A few types of fabrics, such as felt, bark cloth, and paper, are made from materials where the fibers are processed, but not completely separated, so that no recombination stage is needed. Either the fibers or the fabrics are often dyed, and decorative patterns can be formed both through dyeing and through the processes of fiber combination chosen for fabric production. Basketry, which includes mats, chairs, fish traps, and fences as well as containers, tends be made from more rigid fibers and fabrics than textiles, but otherwise it would be difficult to draw a clear demarcation line between the variations possible within each of these crafts.

Due to their low incidence of preservation, there is much less written on archaeological fiber-based materials than on lithics. However, attention to archaeological fibers has increased dramatically in the past few decades, in part because of technological advances in investigating past traces, but largely due to the increased interest in women's work, domestic crafts, and daily life. As has happened frequently in archaeology, once attention was focused on a topic, methods were found or developed to investigate the previously "unknowable." Scattered clues, overlooked or relegated to notes and appendices, were assembled to form the basis for targeted studies of textile production and consumption. (A similar situation has occurred in the archaeological study of ritual and religion, as discussed in Chapter 6.) Good's (2001) summary article in *Annual Review of Anthropology* is probably the best starting point for an introduction to archaeological investigations of fiber crafts, particularly those in English. She cites recent summary books and edited volumes, such as Barber (1991; 1994), Drooker and Webster (2000), Jørgensen (1992), Seiler-Baldinger (1994), and Walton and Wild (1990), as well as older classics. Other useful places to start are the edited volumes from the Northern European Symposium for Archaeological Textiles (NESAT) in the 1990s, several of which were co-edited by Jørgensen. Wendrich (1999: Ch. 3) reviews and compares over a dozen classification systems in English, French and German for basketry in particular and fiber crafts more generally. Wendrich (1999: 55) notes, for example, that while both Emery (1980) and Seiler-Baldinger (1994) have created worldwide classifications of fiber craft techniques, they are complementary rather than rival systems. Emery's system is centered on the structure of the textile while Seiler-Baldinger focuses on production method. Because this is a book on technology, I will make much greater use of Seiler-Baldinger's book, but any serious student of textile production must refer to Emery as well. Finally, for descriptions of ethnographically and historically known fiber processing, spinning, weaving, and dyeing techniques worldwide, Brown (1987), Hecht (1989), and Hodges (1989 [1976]) provide excellent overviews.

I have always been mystified by the assumption that metal working is the most complex craft, requiring the greatest knowledge. For me, the fiber crafts are by far the most complex and difficult of the major craft groups, with more stages of production, more options, more planning, and more complex three-dimensional thought processes necessary to create finished textiles and basketry objects. Perhaps it is because my own studies have involved high-temperature transformative technologies, so that my firsthand knowledge is strongest in these crafts. Or it may be related to my lack of kinesthetic knowledge about these crafts ("body knowledge" or hands-on experience, as opposed to reading knowledge). I have never tried weaving or basket-making myself, and I know from metal working that very confusing descriptions in the literature become much more clear (or more clearly incorrect!) with even a little practical experience. This is one of the great advantages of exploratory experimental archaeology, as I discussed in Chapter 2.

Elizabeth Barber (1994: 17–23) illustrates this point wonderfully in her description of her attempts to reproduce an ancient fragment of cloth found preserved in the salt mines of Hallstatt, Austria. The cloth was a plaid twill pattern in two colors, green with brown stripes. To briefly summarize, Barber laboriously set up her loom with her sister for the irregular warp (vertical) pattern, and then dealt with the difficulties of a set of regular but very narrow alternating weft (horizontal) stripes. At this point, she suddenly realized that the entire process would have been much faster and easier if she had simply reversed the warp and weft threads. Because the ancient fabric did not have any finished edges preserved, it was impossible to tell from the fabric what the correct direction of warp and weft had been. By attempting the replication of the fabric, however, the directionality became obvious.

Kinesthetic knowledge aside, I suspect that the complexities of the fiber crafts truly are often underestimated, partially because of their relative invisibility in the archaeological record, and partially due to their very long and gradual development. Definite evidence for fiber crafts now goes back to the Upper Paleolithic in the Eastern Hemisphere and to the early inhabitants of the Americas (Soffer, *et al.* 2000; Adovasio cited in Good 2001). The importance of experimental and replicative research for fiber studies has been noted above, but ethnoarchaeological research has been similarly insightful. For example, William Belcher's ethnoarchaeological studies of modern-day village fishermen in Pakistan, including apprenticeship to learn net-making (see Figure 3.16 on page 76) and fishing techniques, led to the identification of the use of fishing nets in the Indus civilization, even though no archaeological examples have been preserved. The existence and use of ancient fishing nets was deduced from characteristic rubbing marks on terracotta net weights which had formerly been classified as large "beads," as well as from a few paintings on pottery and from the size classes of fish remains from archaeological sites (Belcher 1994).

Many archaeologists would also point to the fact that women have typically practiced many of the fiber crafts in most societies, particularly fiber and fabric produced for household-level consumption, and women's work has been greatly understudied in archaeology until recently (Conkey and Gero 1991; Wright 1996a). The increased archaeological interest in women's lives over the past twenty years meshes closely with the time period when there has been an increased attention to archaeological analysis of fiber crafts. (Other disciplines have held these interests for longer.) The high value placed on metals in the traditional chronologies of European archaeology also has played a role; for example, the emphasis on the development of bronze and then iron and steel weapons, and the Industrial Era dependence on iron and steel tools and building materials. If the dominant tradition in archaeology and history today had developed out of the New World civilizations, where metals were important primarily as ornaments and textiles were an extremely important status and tribute item, perhaps we would see a different emphasis on the importance of stages of textile development in human prehistory and history.

The general production of fiber and fiber-based objects employs the following stages (Figure 3.13):

1. Collection of fiber-bearing materials
2. Preliminary processing to clean fibers
   (Materials made without separation and recombination of the fibers, such as felts, bark cloths, and paper all move directly from step (2) to step (4). Good quality silk also does not need spinning, but can be recombined by weaving, etc., after initial processing.)
3. Orientation of fibers and creation of strands/threads; combination of strands to form cordage
   (dyeing of fibers occurs at this point – see (5) below)
4. Ordering and combination of processed fiber strands to form fabric or objects (knotting, looping, weaving, coiling, wrapping, etc.) OR creation of massed fiber objects (felts, bark cloth, paper)
5. Ornamentation and joining (dyeing of fabric, applied ornamentation, joining of fabric pieces).

COLLECTION AND PRELIMINARY PROCESSING OF FIBERS

There are many type of fibers, but they can be generally divided into fibers deriving from animals or from plants. Plant fiber use is of greater antiquity and

Extractive-Reductive Crafts

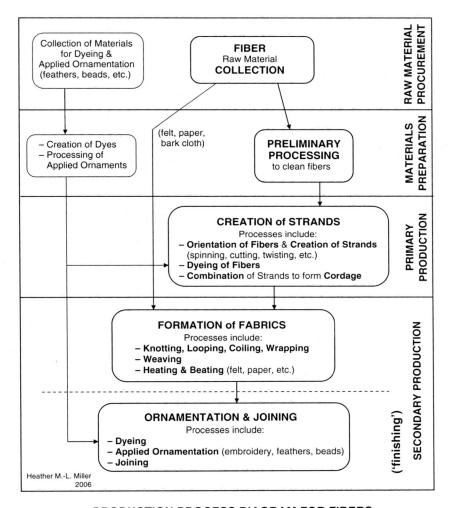

PRODUCTION PROCESS DIAGRAM FOR FIBERS

FIGURE 3.13 Generalized production process diagram for fibers, basketry and textiles (greatly simplified).

broader use, although agricultural societies around the world subsequently developed both domesticated animal and plant species specifically for fiber production, such as sheep, camelids, silkworms, cotton, hemp, and linseed (flax). Plant and animal fibers can be further subdivided into types relating to the part of the plant or animal used.

Animal fibers include hairs, tendons, and insect-produced filaments. Tendons (sinews) were commonly used in situations where very strong, tough, coarse fibers were needed, such as for the assembly of tents or canoes. Specialized filaments produced by the silkworm family (*Bombyx* species and relations) were the source of silks. By far the most commonly used animal fibers, however, were those from hairs. Hairs from almost all types of domesticated animals and many nondomesticated animals have been used, including sheep, goat, dog, cattle, yak, camel, camelid (llama and alpaca), horse, rabbit, and beaver. Wool is a special type of hair from domesticated animals specially bred to produce it, notably varieties of sheep and goat in the Old World and varieties of llama and alpaca in the Americas.

Plant fibers are produced from a very wide variety of plant parts, from the seeds to the roots. Widely-used fibers from seeds or fruits include cotton (from the *Gossypium* species), silk-cotton (from the tree *Bombax ceiba*), and coconut fiber or coir. Leaf fibers are used from a wide range of plants, including the maguey or *Agave* species (discussed below), various palms, banana, and papyrus. Both seed and leaf fibers are particularly common in tropical or semitropical species of plants. Root fibers from the cedar tree were used for basketry and cordage in northwestern North America and elsewhere. Bast fibers from both the stems of annuals and from trees were used around the world, in almost every type of environment and society. Stem bast fibers were used to make flax or linen, hemp, jute, and ramie as well as other products of the nettle family in Asia and Europe. Countless species of whole grasses or reeds were used for cordage and basket-making (see Figure 3.17a on page 77). Among the numerous tree species stripped to make bast fiber or fabric from the bark under-layers were many species of *Ficus*, elm, birch, cedar, willow, and many species of mulberry, including *Broussonetia papyrifera*, the paper mulberry or bark-cloth (*tapa*) tree. Finally, young branches or even older wood from shrubs and trees, especially willow, were processed to create materials for basketry used to make containers, mats, and fences (Figure 3.17b & c on page 77 & 78). Wild or planted shrubs and trees were selectively pruned (coppiced or pollarded) for one or many years prior to the harvesting period to create properly-shaped shoots or branches used for basketry, as well as for wooden objects like tool handles (Seymour 1984; Verdet-Fierz and Verdet-Fierz 1993). These sorts of tended woodlands illustrate the great range of human-plant relationships that exist between "wild" and "domesticated" landscapes.

The collection method typically used to gather animal hair fibers is cutting (as in the shearing of sheep), but loose hair can be combed out or gathered from bushes and thorns, or hair can be plucked out roots and all from the animal. Collection techniques leave other identifiable traces; for example, hair that has been cut for the first time will have a more tapering end than hair cut from an animal previously shorn (Hodges 1989 [1976]: 124). Sinews are

Extractive-Reductive Crafts

gathered as part of the hunting and/or butchery process of animal meat use, while silk fiber cocoons are either gathered from forested areas for the wild insects, or harvested from the containers in which domesticated silkworms are raised. Plant fibers, like animal hairs, tend to be harvested by cutting a section from the rest of the plant, either as leaves, roots, or lengths of bark or stem. A number of stem bast fiber plants are gathered by plucking the plant from the ground, though, notably the flax-producing linseed plant (*Linum*). Seed-borne fibers are gathered from the plant with the seed, as in the case of cotton.

After collection, the processing of almost all fibers involves stages of washing or soaking to clean and help separate fibers from non-fiber materials. Some plant fibers are carefully dried for longer-term storage and only soaked immediately prior to the next stage of processing, as is done for tree and shrub shoots used for basketry. In other cases, the washing and soaking stages occur at once, and the cleaned fibers are stored for future processing (Figure 3.14). In addition, many of the animal and plant fibers are beaten, to soften the materials and make the fibers easier to separate from each other and from nonfiber materials. Frequently, nonfiber materials also had to be more laboriously removed by hand-cleaning or combing. Combing additionally orients the fibers in a similar direction, an advantage if the fibers were going to spun to create a strong thread. Carding, the process of working wool fiber between the short pins or spikes projecting from two flat cards, also cleans out non-fiber material, but it tends to make the fibers fluff out in various directions,

FIGURE 3.14 Tree fibers of *Gnetum gnemon* (Tok Pisin *tu-lip*) hanging to dry after processing by Mandi Diyos (right), prior to thigh-spinning for string bag (bilum) manufacture. See MacKenzie 1991: 70–73 for detailed processing information.

making a fuller but less strong thread (E. W. Barber 1994: 22; Fannin 1970). Hand-cleaning include such processes as the picking out of cotton seeds from the fiber, the shredding of leaves to release fibers, and the peeling of shoots.

Once the fibers are clean, the production of strand production can begin. Most textiles and basketry first produce oriented strands of fibers, then recombine them into a fabric, although there are some fabrics that are made directly from massed, unoriented fibers without separation and recombination of the fibers, such as felts, bark cloths, and paper. These latter materials all move directly from the stage of fiber collection and preparation to fabric production, skipping the stages of strand production. They can thus employ even very short fibers, since the fibers do not need to be long enough to be spun into thread or directly woven into fabric. The processing of skins to make leather or furs is a completely separate process from fiber and fabric processing, although strips of leather can also be used as strands for production of cordage or woven fabric. While I cannot address the important craft of hide production in this volume, the recently published edited volume by Frink and Weedman (2005) provides an entry into this literature through a focus on gender aspects of hideworking.

Production of Strands and Cordage

Threads are the long, organized fibers created by spinning or twisting. I will adopt Wendrich's (1999: 27) use of the term *strand* as a more general term to include both spun threads and the organized but un-spun fibers produced for basketry. The first step in strand production of any type is the orientation of fibers. The cleaning stage may have already done this, particularly if combing techniques were employed. For the production of baskets, the various coarse plant fabrics might need to be further shredded, as in the slitting of shoots for basketry. The organized strands are kept in order by wrapping or coiling or sorting, depending on their flexibility and size.

Once oriented in an organized fashion, some fibers could be used directly, such as many of the strands used in basketry. High quality silk filaments also do not need to be spun, but forms long fibers which can be directly combined by weaving or other processing after initial processing (Hodges 1989 [1976]: 125). Most fibers used to make textiles are too short to be formed directly into fabric, however, so must first be combined into long strands by twisting or spinning. Some basketry fibers must also be combined, as in the grasses wrapped or plaited to make coiled basketry.

Fibers used for thread production thus must be long enough to be spun by one method or another. Spinning is usually divided into two types of methods, hand twisting and use of a spindle, although sometimes only the second method is described as "spinning." In hand twisting or thigh spinning, the fibers are twisted between the hands or between the hand and a flat surface

Extractive-Reductive Crafts

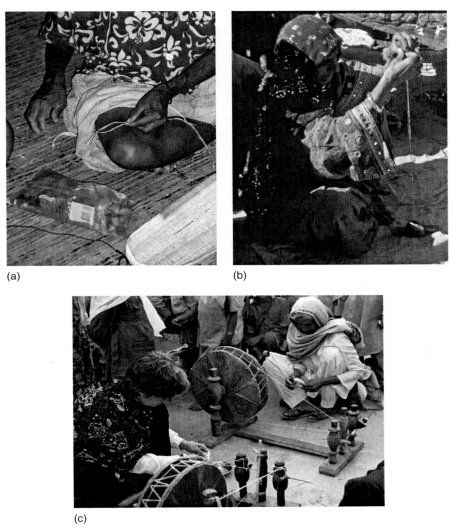

(a)　　　　　　　　　　　　　(b)

(c)

FIGURE 3.15 Making thread by (a) rolling *tu-lip* fibers on thigh; (b) spinning wool with a drop spindle; (c) using a spinning wheel with cotton fiber.

such as a thigh (Figure 3.15a). Additional fibers are fed into the twist at intervals to sustain the growing strand. This method generally requires fairly long fibers, such as stem or bark basts (Figure 3.14), while spinning with a spindle or spinning wheel works well for short fibers, which are the majority of the materials used for woven textiles (Seiler-Baldinger 1994). The use of a spindle employs the additional rotary force provided by the spindle's weight to create a tighter, more even, and therefore stronger thread. Spindles were used worldwide, and in their most basic form consist of a disk-shaped whorl pierced by a rod or shaft and mounted near one end of the rod. The rod is usually of wood, while the whorls are made from a diversity of materials, from wood to reshaped potsherds to carved stone to metal. The fiber is attached to the end of the rod away from the whorl, and the spindle spun until the fiber has a tight twist, creating a strand or thread. This length of twisted fiber is then wound around the spindle, a new length of fiber unrolled or pulled out from a mass, and the twisting process is repeated. (See Brown 1987 for excellent descriptions and illustrations of this process.) Drop spinning involves dropping the spindle from a height during the process of twisting, to impart extra twist (Figure 3.15b). Some fibers resist twisting more than others, such as heavy bast fibers like maguey, and so require harder spinning than fibers that twist more naturally, such as cotton or wool. The spinner (or spinster, to use the older term) can produce a harder twist by imparting more force to the spin, by using a heavier whorl, and/or by using drop spinning. The spinning wheel also imparts greater rotary force, in the simplest cases by adding the rotary motion of the spinning (drive) wheel to the motion of the spindle (Figure 3.15c) (Fannin 1970; Brown 1987).

Animal hairs are covered with scales or bracts, which become interlocked during spinning and help maintain the twist. The attributes of these bracts are one of the clues used by specialists to identify fibers in unknown textiles, although DNA test can also now be used if actual fibers, as opposed to casts or pseudomorphs, are preserved (Good 2001). Mature cotton fibers aid in maintaining twist through their tendency to corkscrew, but immature or overripe fibers are either too limp or too stiff (Hodges 1989 [1976]). Fibers of different sorts (different animal hairs, etc.) can be mixed together prior to or during spinning, resulting in well-mixed to patchy distributions of the different types. The types used have to blend well, however, or the resulting thread might not have the desired properties.

Spun strands might be twisted or spun together to form thicker strands, called a ply or yarn. Typically, the direction of the ply will be opposite that in which the strands were spun, to prevent untwisting. The direction in which the thread or ply is spun, toward or away from the body for thigh twisting, clockwise or counterclockwise for spindles, will determine the direction of the strand's twist. The terms used to describe the two types of twist, Z and S, come from the direction of the spirals of the thread, which look like the angles of the letters Z or S. The direction of the spin has been used to examine the right- or left-handedness of the spinner, the type of thread usually spun, methods of spinning used, and cultural patterns of taught methods of twisting. A great deal of useful information has come from such investigations, although the researcher must be cautious due to the complexity of alternative explanations

(Good 2001; Hodges 1989 [1976]). Cordage is typically created by twisting together two or more strands made from various fibers. Cordage can also be created by plaiting or braiding, but this method is typically used only for relatively short lengths. The finished strands or cordages used for both textiles and basketry are frequently dyed prior to fabric creation. Given the similarities in the process of dyeing at both stages, all dyeing is discussed in one section following the description of fabric production.

For some unwoven fabrics involving the looping of continuous strands, strands can be made directly into fabric as soon as they are produced, with an alternating rhythm of spinning a sizable length of strand, looping a section of fabric until the strand is exhausted, then creating more strand. This is the process used in string bag production in New Guinea, as described fully by MacKenzie (1991). More often, dyed or undyed strands are stored in bundles, coils, or some other organized fashion that will allow easy and even unrolling during the process of fabric creation.

Fabric Production

As noted above, some fabrics are produced without separation and recombination of the fibers. Bark used for bark cloth is soaked and beaten, and natural gums in the bark hold the cloth together, unless it is soaked again. Paper made from papyrus leaf followed a similar process, as did other early types of paper. However, the more common pulp methods of paper-making creates a more even surface, and larger quantities. Plant materials are soaked, beaten, then macerated and soaked again. This pulpy mass would then be spread out over a "wove" mould, a frame holding a porous material such as fine cloth, and the excess moisture drained (Hunter 1978 [1947]). Alternatively, a "laid" mold made of thin grass or bamboo matting could be dipped into the container of pulp. The paper sheet was then removed from the mold, and further dried ("couched") and flattened by various methods. Finally, felt is made from animal hair by washing the wool or fur, laying it out in a layer, beating it ("fulling"), dampening it, and heating it to steam it. The beating and heating steps are repeated as necessary. Alternative processes for felt-making do not involve heat or beating, but simply rolling and unrolling (Hodges 1989 [1976]: 131).

Typically, however, fabric production requires the ordering and combination of processed fiber strands to form "fabric" in the form of cloth, mats, bags, hats, baskets, and basketry fences. The combination of strands employs an enormous range of processes including knotting, looping, coiling, wrapping, and weaving. Given the emphasis on weaving in present and past societies, I will divide the examination of fabric production processes into woven and

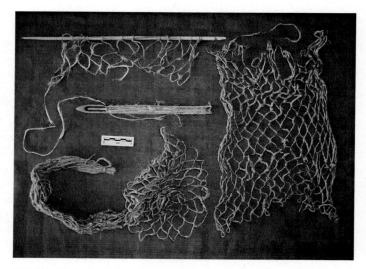

FIGURE 3.16   Examples of net-making tools and portions of fishing nets.

unwoven fabrics, and focus on the former. This is not necessarily reflective of their ubiquity of use, however, and unwoven techniques are likely much older, even for spun fabrics. Seiler-Baldinger (1994) provides a more extensive discussion of techniques organized in a much more exhaustive fashion. Also see Wendrich (1999: Chapter 3) for a discussion of the different types of classification systems that have been proposed, including Seiler-Baldinger's.

Unwoven fabrics employ knotting, looping, coiling, wrapping, and other techniques to combine fibers. *Knotting* is primarily used to make nets, from large coarse fishnets to delicate decorative lace and cloth edgings (Figure 3.16). Knotting systems are often designed to allow the use of continuous strands, to make large pieces of fabric. *Looping* systems are also often designed for continuous strands, where new strands are added on or made by twisting more fiber, as in the making of string bags in New Guinea cited above. Knitting and crocheting are two well-known looping systems. *Coiled* systems are common in basketry, where a strand or bundle of fiber is coiled in a flat or three-dimentional circular pattern (Figure 3.17a). Coiled materials are most often secured by wrapping or sewing the rows of coiling, sometimes making more complex patterns of the wrappings. The rectangular basket in Figure 3.17b is finished at the top border with wrapping around a bundle of strands, and the handles were also made by wrapping.

Woven fabrics, including mats, baskets, and textiles, are made from two sets of strands, running approximately perpendicular to each other, which are woven over and under each other in various patterns. The most common

Extractive-Reductive Crafts 77

(a)

(b)

FIGURE 3.17 (a) Coiled basket made of grass; (b) Woven rectangular basket made from dyed wood strips.

systems have one set of strands which are *passive*, fixed more or less in place, while the other set is *active*, moving over and under the passive set at right angles to it. (Seiler-Baldinger (1994) and Wendrich (1999) describe other types of systems as well.) Woven basketry fences were made by anchoring upright stakes or rods at intervals (the passive system) and weaving strands made from shoots or split timber in and out between the uprights. In the

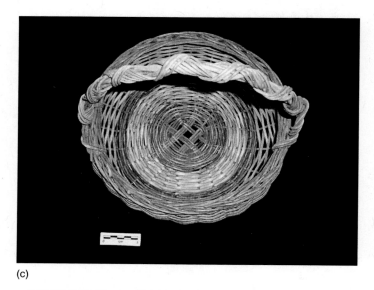

(c)

FIGURE 3.17 (*Continued*) (c) Woven circular basket made from twigs.

rectangular woven basket shown in Figure 3.17b, the base was made by weaving together two sets of active strands at right angles to each other, then turning up these strands to form the passive system through which the active strands (thinner strips of tan, brown, and greenish-blue) were woven. In the round basket shown in Figure 3.17c, the passive strands are laid over and under each other once in the center, then spread out like the spokes of a wheel to allow the active strands to pass over and under them, first in bundles of two or three strands, and then as individual strands. The primary tool used to weave basketry are the fingers and hands of the weaver; other tools are relatively simple, such as a knife or sharp edge for cutting, sometimes a pointed tool or spatula for separating the passive strands and easing in the active strands, and for some materials a rod for "beating down the weave," that is, compressing the active strands by hitting them.

For textiles, made of less rigid materials, weaving is enormously simplified by tying the passive strands to a support, so that they stay in order and do not need to be reorganized with every pass of the active strand. If the passive strands can be made somewhat rigid by placing them under tension, it is even easier to manipulate the active strands. All looms essentially fulfill these functions, by providing a frame onto which the passive threads can be mounted and held under tension. Thus, the development of looms is seen as a critical point of technological change in the process of textile production (Good 2001); it is not surprising that looms of one kind or another are found worldwide.

Extractive-Reductive Crafts 79

In looms, the passive strands are called the *warp* and the active strands are called the *weft*. (Figure 3.18) Looms can be characterized by the way they place the passive strands (the warp) under tension, and by the methods of separating the warp to weave the active strands (the weft) (Hodges 1989 [1976]). The three basic methods of creating tension in the warp are (1) use of a fixed end and the human body, (2) use of a fixed end and weights, and (3) use of two fixed ends. The body-tensioned or back-strap looms are primarily used for relatively narrow strips of cloth, as produced on the back-strap looms of Central and South America (Figure 3.18a). These looms are more or less horizontal, with the warp strands spaced out on bars of wood near either end, one end tied to a fixed object like a post or tree, and the opposite end wrapped around or tied to the waist of the weaver. The weaver sits or sometimes stands to weave, adjusting the tension on the warp by adjusting her or his body, and extending out the warp as the weaving progresses. The second type, the warp-weighted loom, is necessarily vertical (upright), with the upper ends of the warp strands hanging from a horizontal bar and the lower ends tied to weights of stone or clay (Figure 3.18b). The weaver stands or sits, depending on how far down the length of cloth he or she has progressed. The variety of loom weights found in Europe testifies to the popularity of this type in ancient times (E. W. Barber 1994), and Hodges (1989 [1976]: 135) discusses the particular advantage of this variable-tensioned loom for unevenly spun thread. The last type of loom, sometimes called the beam-tensioned loom, fixes the warp to bars on both ends, so that the loom can be vertical but is more commonly horizontal, like the Middle Eastern ground looms and most other non-mechanized looms still in use worldwide (Figure 3.18c). The weaver again stands or sits in front of a vertical loom, but usually sits when using a horizontal loom.

The many methods of separating the warp strands to weave the weft (the active strands) through them also testifies to the ingenuity of weavers through time and around the world. The simplest method, of course, is just to move the weft strand back and forth through the warp by hand, wrapping the weft around a long object (a *spool* or *bobbin*) to make it easier to draw the strand through without tangling. Much more efficient is the creation of a *shed* of some kind, a space between the passive (warp) strands through which the active (weft) strands can easily be passed (Figure 3.18b and c). A shed can be created by passing alternate warp strands through a hole on either end of a card of wood or stiff other material, and rotating the card. However, these card sheds are best for relatively narrow fabrics made of light-weight materials. The most common way to create a shed is by passing alternate warp strands behind the *shed rod*, which is pulled to produce a space between the alternate warp strands. In order to pass the weft back through the width of the fabric in the opposite direction, though, the warp strands then need to be

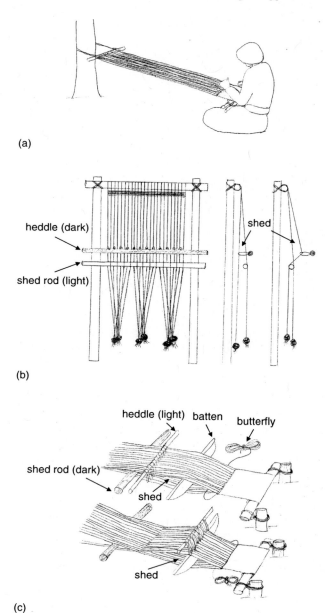

FIGURE 3.18 Drawings of different types of looms: (a) sketch of back-strap loom; (b) drawing of warp-weighted loom, with schematic of shed formed by shed rod and heddle; (c) lower portion of a horizontal beam-tension loom, showing details of shed formed by shed rod and heddle operation. (Redrawn after Brown 1987 and Hodges 1989 [1976].)

Extractive-Reductive Crafts

separated in the opposite direction. Use of a *heddle* allows this, a rod placed in front of the fabric and passed through short loops of thread attached to each of the warp strands that are behind the shed rod (Figure 3.18b and c). Many more complex systems have been developed for separating the warp threads to create the shed, one of the first being the use of a foot peddle to raise the heddle, allowing faster fabric production and more complex woven designs.

Brown (1987: 7–18) outlines the general process of weaving, providing clear descriptions and sketches of the stages and the plethora of specialized tools that have been developed for each of these stages. She begins with the organization of the active (weft) and especially the passive (warp) strands into convenient packages, as discussed in the previous section, and proceeds through the winding, spacing, and tensioning of the warp strands. Brown then discusses the different ways of "making the sheds," that is, providing a way for alternate warp strands to be raised so that the weft can more easily pass through (Figure 3.18c). The next step is the beginning of the actual weaving process, carefully weaving the weft strand through the warp strands if there is no shed, or "throwing" the warp strands through the space made between the warp strands (the shed) using a butterfly or a shuttle. A *butterfly* is a figure-eight shaped bundle of weft strand, convenient for tossing through the space between the separated warp strands (Figure 3.18c). The *shuttle* can be a simple stick, or a more elaborate rounded or boat-shaped object, around which the weft strand has been wound, again to keep the weft strand better organized and easier to throw. The newly woven weft strand must then be beaten down to make a tight weave, using fingers, a comb, or a rounded flat stick called a *batten*. Throughout the weaving process, the width of the fabric needs to be kept constant. Keeping tension on the warp (passive) strands is the primary way this is done, but rigid sticks or other types of stretchers are sometimes used to aid this process. An excellent compliment to Brown's explanation of the process and her helpful sketches is Hecht's (1989) discussion of weaving in eight traditions from around the world, with lovely and informative photographs of the tools and techniques employed in each case.

ORNAMENTATION AND JOINING

Once the fabric is completed, whether basketry or textile, additional ornamentation may be done, or pieces may be joined together to form an object or clothing. Every kind of ornamentation method imaginable has been used, especially for textile decoration, and many are discussed and categorized in the books cited in this Fiber section. Here, I will restrict my comments to dyeing, applied ornamentation, and joining.

As mentioned previously, either the processed fibers (strands) can be dyed prior to fabric production, or the fabric can be dyed after completion. The same dyes are used in either case. Basketry dyeing is almost always done for the strands, not the finished fabric; it is easier for the strands to absorb color while soaking and easier to fit loose strands in a dye pot than to fit a finished basket. Basketry materials are first treated with metallic salts or plant extracts to increase dye absorption before immersing in the dye. As with textiles, dyeing at this stage allows the craftsperson to use the weaving of the colored pieces to make additional patterns.

Dyes are divided into *substantive* dyes, which combine easily with plant and animal fibers in water, and *adjective* dyes, which require the addition of other materials for the dyes to join with the material. Adjective dyes include both *mordant* and *vat* dyes. Mordant dyes, the largest group of dyes, are used with metal salts (mordants) that form a link between the dye and the material. Both substantive and mordant dyes are water soluble, but mordant dyes will not fix to the material without the added mordants. In contrast, *vat* dyes are a small group of dyes that do not dissolve in water, but require special treatment and exposure to oxygen, such as indigo and Tyrean purple. Hecht (1989) discusses all of these dye groups in an overview section, providing a substantial reference list of works on dyes and dyeing, and then explains the procedures for specific dyes and specific fibers for each of her eight world areas.

To elaborate on these categories of dyes, substantive dyes are a relatively small group of natural dyes that are water-soluble and combine easily with plant and animal fibers. Some of these substantive dyes combine permanently with the fibers via a chemical reaction, most notably lichens, which contain colorless acids that serve to fix the dyes without any additions. Other substantive dyes do not combine chemically, but the color is simply absorbed into the fiber and washes out or fades over time. Several yellow dyes are of this type, including the safflower used for the robes of Buddhist monks, and turmeric used for cotton cloth in India. In both of these cases, the fabric would simply be re-dyed periodically. These substantive dyes are applied by crushing the dye source, dissolving it in water, sometimes heating the mixture, and dipping or soaking the processed fibers or fabrics in the dye vats. The fibers or fabrics are usually agitated to ensure even dyeing, by stirring with a stick or even with the feet.

The majority of natural dyes, mordant dyes, are soluble in water but are not readily absorbed by fibers on their own. Mordants chemically bond the dyes with the fibers, resulting in strong and mostly permanent colors. Mordants are metal salts found as mineral deposits and in plant compounds, as well as in urine and edible salt. The mineral mordants include alum (potassium aluminum sulfate), tin (stannous chloride), and ferrous sulfate, while plant

Extractive-Reductive Crafts

compounds used as mordants include tannin, lye (made from wood ash), and vinegar. Again, specific mordants were useful for specific dyes and specific fibers, and some mordants were more damaging to fibers than others. Alum was particularly good for a large range of permanent reds and yellows, and damages the fiber relatively little. Red dyes requiring mordants include madder, made from a plant root (*Rubia tinctoria*) native to Europe, and cochineal, made from an insect living on cactus plants in desert portions of the Americas. (Indeed, the three most popular red dyes in the past were made from insects: kermes in Europe, lac in Asia, and cochineal in the Americas.) Yellow dyes typically come from plants, such as weld (*Reseda luteola*), indigenous to Europe but widely cultivated, which produces very different shades with different mordants.

Finally, the vat dye indigo, from the plant *Indigofera tinctoria*, is one of the rare dyes not soluble in water. The plants are usually first fermented in water to extract the dye source then oxidized by beating with paddles, although other methods are used. The resulting materials are made into dried balls or cakes for trade or storage. The dye source is then dissolved again in a bath containing an alkali, such as lime or potash, and fermented again. Finally, the fibers or fabric are immersed in the dye bath, and on exposure to oxygen turn a deep, permanent blue.

Particular dye sources work best on particular types of fibers, and relatively subtle differences in processing, such as heating the mixture or varying the length of time a mordant is in contact with a fabric, can make all the difference in whether a dye produces a strong, vibrant color with minimal damage to the fiber. Thus, even the simple substantive dyes can require complex, case-specific knowledge. It is not surprising that dyeing was an aspect of fabric production most often practiced by separate specialists, who guarded their dye secrets closely. Control of sources of alum, the favorite European mordant, played a major role in the economic intrigues and rivalries between the great European merchant cities of the Renaissance era (fourteenth through sixteenth centuries AD/CE).

Dyed fabrics can be made very elaborate through the use of resist or applied patterns. Applied patterns could be applied with a brush (painted fabrics) or printed with stamps or blocks dipped in dye compounds (Figure 3.19a and b). Resist dyeing used wax, tied strings, folding, mud, and various pastes to create patterns on fabric, or even on the strands in the case of ikat weaving. The fabric was then immersed or dipped in dye, which was blocked from the patterned areas by the resist material. Removal of the resist leaves un-dyed fabric in the pattern desired (Figure 3.19a, left). Fabrics could be previously dyed, resist patterned, and then re-dyed, or the area where the resist was applied could be shifted between a series of dyeings, to create multicolored fabrics.

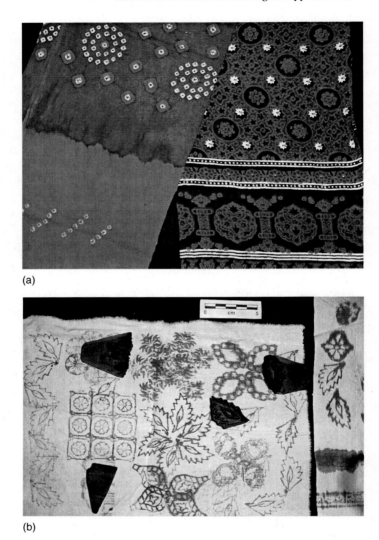

FIGURE 3.19 (a) Examples of tie resist dyeing (left) and block printing (right); (b) example of carved wooden stamps used to print cloth.

Applied ornamentation could create equally colorful effects, through decorative stitching such as embroidery or the attachment of other pieces of fabric cut into shapes. Threads used for decorative stitching are usually plant or animal-derived fibers, but threads made from precious metals are also used. Strands of fiber or metal can also be used to create decorative edging for fabric in the form of looped or knotted borders, lacework, and fringes or tassels. Thick strands of fiber could be pulled through a loosely-woven fabric and knotted or looped to create designs on pile fabrics. The addition of beads or feathers to fabric, usually with fiber thread, has been used to create clothing signaling high status in many societies. Craftspeople in Eastern North America used dyed porcupine quills to decorate some fabrics, although more usually this was done on leather or bark materials. Emery (1980) and Seiler-Baldinger (1994) outline an astounding variety of applied ornamentation. Joining could also be considered a type of ornamentation. For example, the joining of narrow strips of fabric of different colors or fibers is one decorative technique widely used where the back-strap loom is employed. Joining is also used in the construction of complex clothing, which is sometimes practical but is also typically ornamental, and functions as an important means of social communication (Weiner and Schneider 1989).

ORGANIZATION OF PRODUCTION AND SCHEDULING DEMANDS

Fabrics are a major means of social communication, and a major component of domestic and wider economies. As with the beads discussed in the section on Value in Chapter 6, fabrics can be used to convey status differences or group allegiance. Clothing also is used to convey gender and age groupings in most societies, as well as providing protection from environmental extremes. It is thus not surprising that fibers, fabrics, and other materials used in producing them, were widely traded. Collection and production of these materials formed a fundamental component of the economies of all societies, whether created for a family as part of the work of each household or produced on a large scale by full-time specialists (Brumfiel 1991; Costin 1996, 1998a; Sinopoli 2003; Weiner and Schneider 1989; Wright 1996b).

Because fiber production has so often been a household-based economy, but with society-wide economic impacts, the organization of production for fiber crafts provides some excellent insights into the complexity of integrating the various tasks needed to sustain a family or a society. Such scheduling issues have been examined by several studies of textile production in societies where this was primarily women's work, discussing the probable conflicts between fiber production and other work traditionally done by women, particularly food production.

For example, Brumfiel (1991) discusses the increasing demands of textile production for tribute on Aztec households, and how this would have affected the time women had to spend on household tasks. (See the section in Chapter 5 on labor for more on specialized women weavers.) She suggests that some women near the urban regions may have turned to commercial food production for sale in the marketplace, buying their tribute cloth rather than making it themselves. This conclusion is based on debated estimates of spindle whorls (Costin 1991), but there is no question that increasing tribute demands would have placed pressures on weavers, primarily women in the domestic sphere. Brumfiel also discusses variation in the relative proportions of methods of food production in different regions, contrasting the more time-consuming preparation of tortillas, as represented by griddles, instead of stews and atole, a sort of maize porridge, as represented by pots. Again, while the methodology could be debated, Brumfiel's attention to the varying scheduling constraints involved with food preparation encourages contemplation of these issues. One of her most important points is that there were multiple methods of solving the problems of conflicting time demands with increasing pressures in one or more spheres of household production—a point that is more commonly made for extra-household craft production, in discussions of specialization. Costin (1996) similarly notes that there were greatly increased demands on Andean women after the Inkan conquest, in order to supply cloth for the cloth tax as well as to continue to provide cloth for their own families. However, she found no evidence for any corresponding decrease in any of their other duties, and so concludes that women simply had to work harder after the conquest, a conclusion that resonates with the increased work loads of modern families.

A different aspect of scheduling conflict is seen in the need to allot agricultural land and labor to the production of domesticated plant and animal sources of fiber, rather than production of food. In most agricultural societies the majority of clothing and other textiles have been made from domesticated plants and animals. Furs and leathers from wild animals and wild plant fibers are still very important components of the economy, used for ropes, basketry, coats, footwear, and hats in a range of societies and climates. But for societies where cotton, linen, wool, and other domesticated fibers were and are important, the production of these fibers can affect the entire farming system. As part of the overall agricultural system, choices have to be made about using land for food versus fiber crops. The time and labor devoted to the upkeep and management of wool-producing domesticated animals takes away from food production, including the collection and storage of fodder for year-round animal maintenance in cold climates, something not necessary for meat-producing animals that can be slaughtered as fodder grows scarce. To understand the full ramifications of the fiber crafts within a food producing society, it is necessary to look at the entire agricultural system—food, fiber and other production—to understand the choices being made.

An outstanding example of the role of fiber production in agricultural systems is the Parsons' thorough ethnographic and archaeological study of maguey production in the highlands of central Mexico, and highland Mesoamerica more broadly (Parsons and Parsons 1990). While there are more than a hundred species of *Agave* found throughout Mesoamerica and the southwestern United States, primarily desert plants with thick fleshy spiked leaves and tall flower stalks, only a few species have been domesticated. The domesticates are relatively large plants well-adapted to semi-arid highland environments in Mesoamerica, and these domesticated *Agave* species are the plants referred to as maguey by the sixteenth century Spanish and later writers. The Parsons examined all aspects of maguey production, for food, drink, and fiber produced from the flesh, sap, and fiber of this tough, productive plant. For the past several thousand years, prior to and throughout the development of agriculture in this region, maguey flesh was cooked and eaten and maguey fiber processed and used. Historically, these plants were also important for the mildly alcoholic drink *pulque* made from their sap. The Parsons list the many uses of this versatile plant: cooked edible flesh, *pulque* as well as syrup and sugar from the sap, fiber for cloth and cordage, fuel, and materials for construction, to name only the major products. It is not surprising that it was a staple crop in highland Mesoamerica, given its unusual status as both food and fiber plant.

In terms of land use, maguey complements rather than competes with other food plants for agricultural field space on two counts, as it not only grew where other plants such as maize, beans, and cotton would not, but was also inter-planted with other crops or planted on plot edges, providing food in the agricultural off-season, and providing extra insurance for bad years, when only the maguey would withstand cold, aridity, hail, or other disasters. This inter- or edge-cropped maguey also helped prevent sheet erosion, stabilizing the farmland, and the leaves and stalk left after processing was used for roofing, building construction, and fuel. The latter use of maguey should not be under-estimated for this semi-arid environment, especially once urban societies develop. The Parsons note that sixteenth-century accounts specifically mention the sale of maguey stalks for fuel in urban marketplaces (Parsons and Parsons 1990: 365). While the maguey plant has multiple uses, the Parsons point out that using the same plant for all purposes lowers its productivity overall; plants with sap extracted are more difficult to process for fiber, and the flesh is no longer fit for cooking. Therefore, choices had to be made about the particular use to which plants would be put—would the farmer focus on fiber production or would sap production be more useful? Or would both be done, with less sap extracted? The Parsons suggest that particular species and varieties of the maguey plants may have developed by selection not only for microclimates, but also for specific uses.

Interestingly, the Parsons compare maguey's place in the highland Mesoamerican agricultural system not to maize or cotton, or any other food or fiber plant, but to the domesticated camelid (*llama*) of the Andean highlands. (See Dransart (2002) for an ethnographic and archaeological account of llama raising and fiber production in the Andes.) Both the llama and maguey are sources of relatively nonseasonal yet nonstorable food, with the exception of dried meat and maguey sugar respectively, and both are sources of fiber. Both extend the possible areas of food and fiber production into the drier and colder highlands. Finally, as is the case for other sorts of fibers, the spinning and weaving of maguey fibers can take place during the agricultural off-season, avoiding some scheduling conflicts for the available labor.

Nevertheless, maguey plants did require labor input for planting, tending, harvesting, and especially for processing for fiber and food products. This labor was time and energy that could not be spent on other food or craft production tasks. For people living in this environment, the reliability of the maguey was well worth the effort, although at least for modern people, maize and beans are more culturally important crops. The Parsons point out this paradox of the great cultural importance placed on maize and bean crops, which did poorly more often than not, in contrast to the actual economic importance of the maguey crop. They note, however, that maguey may not have been sustaining enough to support people without maize and beans, so that the latter may have been pivotal rather than staple crops in the past. Today the products of the maguey crop are frequently exchanged for needed maize, beans, and other supplies in the market, so that the modern and historic maguey growers were actually specialized producers, a situation that likely existed during prehistoric periods only when these regions were part of complex societies. Nopal cactus (*Optuntia* sp.) was another plant which would have been of great importance for a specialized economy in this region, as the primary food of the cochineal insect, one of the three most important red dyes of the ancient world. It too was a food plant, but one which was only supplementary, grown for its seasonal fruits and possibly the cooked flesh of its leaves.

The Parsons suggest that in early prehistoric periods, maguey was pivotal in allowing the expansion of farming into marginal land, especially in the highlands, although complete dependence on these areas would only be possible once some sort of large-scale redistributive system was in existence, to allow exchange of maguey and other products for needed foodstuffs. Much later, in the Postclassic period (the eleventh through sixteenth centuries AD/CE), maguey was important due to the great demand for cloth made from various grades of maguey fiber as well as cotton, for market exchange and for tribute or taxes. As noted, the usefulness of maguey fiber, fuel, and sap in providing products for exchange continued on into the historic and modern periods.

The fiber crafts themselves supported a number of other crafts working with perishable materials, due to their need for many specialized tools of wood, bone, and antler. Such tools included picks or awls for basketry and bobbins, battens, beater combs, and the looms themselves for textiles. These sculpted organic crafts are discussed in the next section.

WOOD, BONE, AND OTHER SCULPTED ORGANICS (ANTLER, HORN, IVORY, SHELL)

As discussed in the Fiber section above, simple wooden or bone objects employed as digging sticks or clubs may have been the earliest artifacts created and used by early human ancestors, long before the earliest preserved artifacts, stone tools. Nevertheless, the addition of stone tools to the repertoire of our ancestors must have greatly increased their ability to cut and shape larger pieces of wood, as well as bone, ivory, antler, and shell. Tools made from rodent teeth have also been used in different parts of the world for carving designs into objects made from these materials. Later, many metal tools were utilized in sculpting and decorating wood and other hard organic materials, tools such as knives, axes, adzes, saws, planes, chisels, and eventually the lathe. As with the fiber crafts, however, the poor preservation of sculpted organics makes them a challenge to investigate. This is particularly unfortunate given the great richness of woodworking seen in many parts of the world in historic periods.

Ironically, some of the areas with the best preservation of ancient wooden objects, due to aridity, are areas that were particularly poor in wood in the past for the same environmental reasons. Egypt is such a case, but fortunately the very richness of sculpted organic remains and the availability of written records has shown the importance of the timber trade in the past, both with the eastern Mediterranean coast and with other parts of Africa (Killen 1994). Areas with waterlogged and peat-preserved wooden objects, such as northern Europe and the British Isles, also provide evidence for the importance of wood in past artifact assemblages. In this section, I will focus on woodworking due to its predominance among the sculpted organics, but will reference other materials throughout the discussion of processing.

It is noteworthy that production discussions of wooden objects are seldom described as part of a material type (woodworking), and are more often grouped by function: boatbuilding, furniture-making, building construction, weaving tools. Perhaps this is because in those rare cases where evidence for woodworking is preserved, it is often clear what kind of objects were produced, unlike many other crafts. Ivory, shell, bone, and antler, all of which are more likely to be preserved than wood, are more commonly grouped by material type when production processes are discussed, like the other materials described in this book. This has meant that creating a section on woodworking as a production process group has been surprisingly difficult, and there are few general reference texts. Hodges (1989 [1976]) has a strong section on woodworking, heavily focused on the working of wood in European traditions. There are several Shire series on woodworking in particular places, both in the Ethnography series (e.g., Craig 1988) and in the archaeology series (e.g., Killen 1994 in the Egyptology series). It is not surprising to find that the International Council for Archaeozoology (ICAZ) has a Worked Bone Research Group, given that it is usually zooarchaeologists who end up with bits and pieces of worked bone, antler, ivory, and other teeth, often classified together as "bone-working." A recent edited volume from this group (Choyke and Bartosiewicz 2001), with abstracts in English, French, and German, contains a wealth of articles on all aspects of osseous material production and use, primarily from Europe. The editors also helpfully note a number of bone working research groups and published references in their introduction. LeMoine's (2001) article in this volume provides an overview of the literature and a good bibliography for the production of objects from skeletal materials, particularly for North American contexts. The most frequently cited single work on production of objects from skeletal materials is probably MacGregor (1985), indicating the continued strong interest in these objects in European contexts after the Roman period. Researchers working on earlier periods also cite MacGregor for his excellent descriptions and illustrations of the production processes for bone, antler, ivory and horn working, as well as his explanations of the mechanical properties of these raw materials. For Paleolithic contexts and some Neolithic examples in Europe and Western Asia, Piel-Desruisseaux (1998) provides clear descriptions and illustrations of bone and antler working for the production of tools and also handles for stone tools.

The general production of sculpted organic objects, particularly wood-working, employs the following stages (Figure 3.20):

1. Collection of material (selection of wood, shell, horn, etc.; removal from location)
2. Preliminary processing of material (initial trimming and seasoning of wood; removal of animal and bleaching of shell; trimming of antler; etc.)
3. Shaping of objects, employing
   a. Reduction: cutting, including sawing, scraping, drilling, etc.; splitting; or percussion
   b. Alteration: bending, usually with heat and/or moisture and
   c. Combination or Joining by lashing, mortising, gluing, pegging, nailing, etc.
4. Finishing stages, such as abrasion for smoothing or polishing; cutting for engraving and inlaying; hardening; and coloration.

Extractive-Reductive Crafts 91

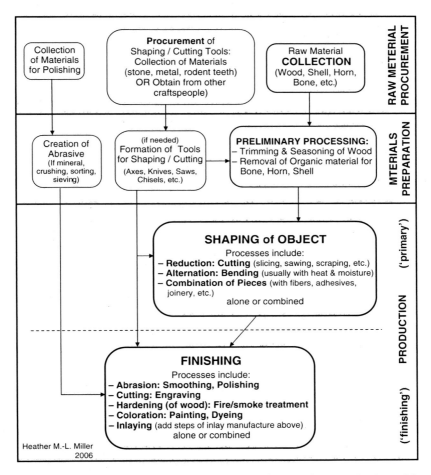

PRODUCTION PROCESS DIAGRAM FOR WOOD & OTHER ORGANICS

FIGURE 3.20 Generalized production process diagram for wood and other sculpted organics (greatly simplified).

COLLECTION AND PRELIMINARY PROCESSING

Selection of wood type for a particular use is a complex and exact process. Particular woods have certain uses based on their properties, as was seen for stone. Ash is elastic but tough and so makes good tool handles and bows, although yew was preferred for bows where available; birch is tough and fine-grained, so useful for turning on a lathe; ebony and boxwood have very dense, fine-grains, and so are ideal for carving. Different wood types are similarly rated for their use as firewood under different circumstances.

Hodges (1989 [1976]) points out that there are three general approaches to wood harvesting. All of the wood can be collected, and then the land can either be allowed to regenerate woodland or converted to farmland and pasturage. Alternatively, the clear-cut area can be re-planted with young trees, as is done in modern times. Finally, the woodland can be managed, with selective trimming of parts of the trees or thinning out of younger trees. This last method was useful in cases of long-term use of an area and has been well-attested for European contexts, especially in the form of coppicing and pollarding, as discussed in the basketry section above and the section on shaping below. Typically, wood was cut or trees were felled with an axe or adze of ground stone or metal; saws were only employed at relatively late dates, with special systems to prevent the saw from becoming stuck in the trunk ("bound") as the tree settled into the cut. If an entire tree was cut down, the branches would be cut off from the trunk, to be used themselves for wood or for fuel. The bark might also be used for dyeing or tanning, depending on the tree species; bark used for flooring and lining walls is shown in Figure 3.12. The trunk might be roughly squared and used as a large post or log in building, or it would be split, cut, or sawn into smaller posts or planks using one of several methods depending on whether the planks were split along the grain or cut or sawn across the entire trunk (Hodges 1989 [1976]; Killen 1994). Choices about how to harvest and process wood depend on the type of trees, the technology for cutting, the intended use of the wood, and on cultural beliefs and practices. For example, an unusual type of harvesting was practiced on the Northwest Coast of North America, where planks of two to fifteen meters in length were harvested from enormous living cedar trees by two or more people (Figure 3.21) (Stewart 1984). After notching the tree with a stone adze, antler or wooden wedges were pounded in with stone hammers to carefully split off the planks (Archaeology Branch 2001; Terence Clark n.d. and personal communication). Hundreds of such culturally modified, living trees can still be seen in the forests of British Columbia today.

At some point, wood has to be seasoned to avoid warping and splitting. "Seasoning" primarily refers to the slow drying of wood by exposure to the air after the wood is cut, usually after the trunk or branch is split into lengths or cut into planks. Seasoning might take a few months or more than a year, depending on the climate, the type of wood, and whether the wood was to be used indoors or out. During seasoning the wood needs to be exposed to air at a relatively stable temperature and humidity to allow shrinkage yet prevent excessive warping. Early Euro-American settlers in northeastern North America cleverly dealt with this problem by laying their pine floor planks in place, but not nailing them down. After a year of *in situ* seasoning,

Extractive-Reductive Crafts   93

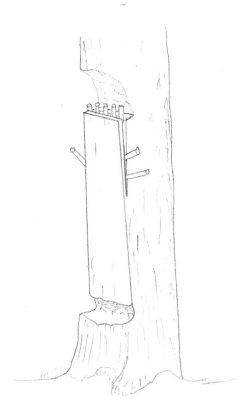

FIGURE 3.21 Living cedar trees from Northwest Coast of North America (British Columbia) with planks removed. (Redrawn after Stewart 1984.)

the planks would be tightly fitted, then nailed in place (Sloane 1962). Planks cut radially through the center of the trunk will warp less than planks cut tangentially from the edges of the trunk, as the wood grain lines (the rings or cells of the wood) of tangentially-cut wood are cut at different angles and shrink at different rates.

Bone, shell, antler, horn, and ivory all had to be thoroughly cleaned after collection, to remove any soft tissues that would rot and smell. Not surprisingly, this process was often done at a distance from working and living sites, particularly if materials were processed on a relatively large scale. Antlers are shed annually by the various deer species, so that they can be collected after shedding; the remaining materials had be collected by killing the animal involved. Ivory and shell would be ready for shaping after cleaning.

Antler, which was relatively hard and dense, might be softened by soaking in water to make shaping easier. Most often, the bones used for object production were either the long bones or shoulder bones of animals. Long bones would be broken, cut, or sawn to remove the ends and marrow, and then further shaped depending on the object desired.

SHAPING AND FINISHING METHODS

The shaping of sculpted organic objects involved reduction methods, alteration techniques, and/or recombination of pieces (Figure 3.20). *Reduction* methods used in shaping these materials usually involved various types of cutting. To echo the definition in the section on stone, *cutting* is the shaping of a material using a tool that is much harder than the material itself, so that the tool can cut the material without use of an abrasive, and without much wear on the tool. This includes cutting, sawing, drilling, and groove-and-snap techniques with stone or metal tools, all as described in the Stone section above. For wood, the most common tools used are axes, adzes, or knives (Figure 3.22). *Shaving* or *scraping* are methods not commonly used in stone working, but very commonly used to shape wood or other hard organic materials by removing thin sheets of material from the surface; tools employed include chisels, adzes, knives, and more specialized tools of stone, metal, bone, or even shell. Turning on a lathe is basically shaving or scraping with a knife, scraper, or chisel of metal while the material to be worked is rotated, and was used for most of the

FIGURE 3.22 Bretaro shaping a wooden door board using an axe. *Photo courtesy of Roger Ivar Lohmann.*

sculpted organics wherever the technique was known. Hodges (1989 [1976]), Seymour (1984), and Sloane (1962) all provide more detailed discussion and illustration of the enormous variety of woodworking tools commonly used in the European and Euro-American tradition. Woodworking also made use of an unusual reduction shaping technique for certain objects, shaping by *burning*, famously used to hollow out logs to create small boats. Bone working also made use of cutting with knives and saws. *Groove-and-snap* techniques were common in bone working, so much so that bone working might be grouped with the production of soft stone objects more readily than with wood and other hard organics. It is not surprising that studies of bone working frequently draw parallels with stone working. Bone working, and sometimes shell working, often employed rough "knapping" techniques in the early stages of shaping, using *percussion* to break open long bones or remove their ends, or in the case of shell to fracture the shells as desired. "Notching" of bone materials, percussion resulting in a cut or notch in the material, could even be done with bone or antler tools (Provenzano 2001; Schibler 2001). The use of wooden or metal wedges hit with hammers to split wood was a similar percussive technique, and was possibly even more widespread than cutting methods for creating planks, as illustrated in the Northwest Coast example in Figure 3.21. Finally, abrasion was sometimes employed in the reduction shaping of sculpted organics, although it was much more common in the finishing stages for smoothing and polishing.

Shaping by *alteration* offers an excellent example of the nonlinearity of even the most basic production processes, for there are cases where wooden artifacts are actually shaped *prior* to collection. The European practice of altering the growing patterns of trees and shrubs to produce desired materials has already been mentioned in the Fiber section above, where pruning of trees and shrubs (coppicing and pollarding) was used to create numerous straight, thin rods for basketry production. The process of wooden handle production for agricultural tools went even farther, with branches of trees of the desired wood bent to create the desired shape of the final handle, as much as a decade prior to the harvesting of the wood (Seymour 1984). The more common method of alteration involved bending of wood using heat and moisture, by holding wet wood over a fire or steaming them in some other way. This technique was widely used for straightening warped pieces of wood and/or bending them to desired shapes. Horn was quite often soaked or boiled in hot water, to separate the growth layers into thin translucent sheets for windows or lantern shields (Hodges 1989 [1976]), or to de-laminate or bend the horn into desired shapes.

Combination or joining methods are as diverse as reductive shaping techniques, particularly for woodworking. Lashing with fibers is likely the earliest method of joining of wood and other sculpted organics, and is still widely

FIGURE 3.23 Lashed pole suspension bridge over the Fu River, Papua New Guinea.

used; Figure 3.11 (above) shows a stone adze hafted to a wooden handle with lashed fibers, while Figure 3.23 shows a suspension bridge made out of lashed wooden poles and houses made of lashed poles are visible in the background of Figure 3.22. A variation on lashing is sewing, where holes are drilled in each piece of wood and fibers or thongs passed through them. Sewing of planks and other materials has been widely used in boat-making, as illustrated in Chapter 5. Adhesives are also a common method of joining, either alone or in combination with other joining methods. Various systems of interlocking notches and projections from mitres to dovetails have been used to combine pieces of wood in "joints," alone or together with adhesives, wooden pegs, or metal nails. Similar techniques have been used to make ivory boxes and other complex pieces. Pegs and nails are also used in combining wooden pieces without joinery systems, as in the early "plywood" found in ancient Egypt, where thin sheets of wood were laminated with their grains at right angles to each other and joined with wooden pegs (Killen 1994:9).

There is a great deal of overlap between what I have designated as "finishing" and "shaping" stages, but in general, finishing stages employ primarily abrasion for smoothing and polishing; cutting for engraving and inlaying; and different methods of coloration. If a wooden object needs to be hardened, as in the production of spears or digging sticks, this is also done after shaping.

As with stone and metal objects, abrasion of hard organic objects can be done with a wide variety of materials, from stone to high-silica leaves (Figure 3.24) to cloth. Quite often stages of abrasion with progressively finer

FIGURE 3.24 Belok abrading and smoothing a wooden doorboard with leaves. *Photo courtesy of Roger Ivar Lohmann.*

materials are employed to create a highly polished surface. The abrasion methods employed for stone are also used for hard organics. Shell beads can be polished in the same way as soft stone beads, for example, strung and rolled along a grinding stone (Figure 3.9, above); in fact the entire production process for hard organic beads is often the same as for the production of soft stone beads. Cutting is done at the finishing stage primarily in the form of engraving, a very common method of decoration for hard organic materials of all types. Engraving tools are found from the Paleolithic, in the form of stone burins, and chisels or gravers made of stone, metal, bone, and shell are used for engraving around the world. Wood, and perhaps other materials, have also been engraved with animal tooth tools, especially rodent incisors (Figure 3.25) (Craig 1988: Figure 31). Engraved designs can also be accentuated by filling them with pigments or by inlaying contrasting materials such as other woods, shell, stone, or metals. This is one form of coloration; others include the use of stain or paint to change the color of the surface or to make designs on it. Plaster or clay of different colors can be applied to wooden objects either to color the surface or to provide a base onto which paint is applied.

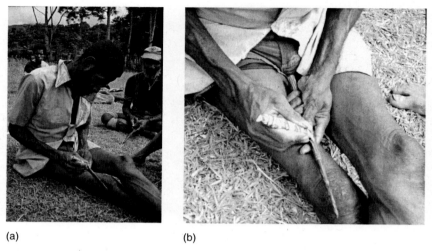

(a) (b)

FIGURE 3.25 (a) Wuniod using a marsupial rodent incisor tool to carve a wooden arrow tip; (b) close-up. *Photo courtesy of Roger Ivar Lohmann.*

ORGANIZATION OF PRODUCTION; USE AND REUSE OF HARD ORGANIC OBJECTS

Reconstruction of production processes are difficult for these perishable materials, particularly as many of the tools used are also made from perishable materials. The discussion of boat building in Chapter 5 illustrates the range of available data types for different regions and time periods. In Egypt, whole wooden boats were preserved, but in the Arabian Sea, reconstructions of boats have been created primarily from impressions in bitumen coatings together with logical reasoning based on the physical forces and stresses involved (Vosmer 2000). In Southern California, little archaeological material is available but an enormous wealth of data has come from ethnographic accounts and experimental reconstructions (Gamble 2002; Hudson, *et al.* 1978; Hudson and Blackburn 1982). In spite of these differences in preservation, the archaeologists in all these cases have done their best to move beyond reconstruction of production methods (a difficult enough problem!) to an examination of the entire technological system, investigating how the production and use of these boats fit into and affected their societies.

In England, work on the Somerset levels and other waterlogged wood assemblages have resulted in quite detailed studies of simple, rough wooden artifacts such as hurdles or trackways. Careful analyses of the type of tools used to cut these artifacts, like the database created by Sands (1997), can be used for reconstructions of the organization of production. For example, archaeologists examine how much of a trackway was made with the same tools, for clues to whether these tracks were created in sections by each community along the way, or whether a single group produced most of the trackways. Analysis of marks on bone tools can similarly provide information on the types of tools used in their production, as in the determination of the types of tools used in animal butchery through cut-mark analysis of bones. Schibler (2001) provides an experimentally-based example illustrating how bone was worked not with either stone or metal but with antler, an unexpected conclusion. In periods where both metal and stone cutting tools are still in use, the relative percentage of metal vs. stone cutting tools used by bone workers might provide clues to their economic status, if other factors can be discounted. This must be done with care, of course; for example, Wake (1999) presents a case where use of metal tools for bone working is not only about economics, as is further discussed in Chapter 5 in the section on Style.

Russell's (2001a) analyses of bone working at Neolithic sites across Eurasia have revealed intriguing contrasts in the organization of production for very similar types of objects, production differences that she ties directly to differences in social organization across the region. She compares the bone points found at sites in southeast Europe, at Çatalhöyük in Turkey, and at Mehrgarh in Baluchistan (the mountainous region forming the boundary between Western Asia and South Asia, between Iran and Pakistan). Using data on raw material selection, production techniques, standardization of products, usewear, and re-use of points, Russell concludes that the value of bone points differs between these locations and also varies across time at each location, albeit with some caveats due to sampling issues. She also concludes from these data that there are significant differences in the organization of labor between Mehrgarh and the sites in southeast Europe, with Çatalhöyük probably similar to southeast Europe based on preliminary results. In southeast Europe, bone production is specialized to some degree, at least at the site level, with some sites apparently focused on bone production for exchange. However, Russell concludes that production continues to remain in the household, and in southeast Europe it is the household structure which is modeled at the center of the social system, bound in gift relations and competing for prestige (not unlike the system modeled for the Northwest Coast of North America). Russell contrasts this bone point production system with that found at Mehrgarh where there seems to be greater craft specialization even during the Neolithic, in terms of standardization of bone tools, as well as spatial differentiation of activities for a number of crafts. She sees the system at Mehrgarh as already showing occupational specialization and a social system oriented towards organic rather than mechanical solidarity (Durkheim 1933 [1893]). As a South Asian specialist, where there is some concern about the very rich ethnographic record overshadowing our reconstructions of ancient societies, it is rather reassuring to see a specialist in the Neolithic of southeast Europe come to many of the same conclusions about the great time depth of occupational specialization and its relation to social organization in South Asian societies.

Determining the use of objects has been an especially difficult problem for scholars studying bone- and antler working, as bone and antler tools are often very simple and could potentially be used in a multiplicity of tasks, frequently tasks associated with perishable materials such as the fiber crafts and hide working. Bone and less commonly antler was used to make tools used in weaving, such as battens and comb beaters, and tools used in sewing, such as awls and needles. Bone and antler tools were used for leather braiding and thong making, and for tools used in basket making such as shaft-straighteners and picks or awls. Scapula were used for hoes. Some bone and antler tools were expedient, but some were clearly valued, as they were decorated, curated, re-sharpened and reused. The re-sharpening and re-use of stone tools has been mentioned in the Stone section. The re-use of wood or wooden objects has been less frequently documented, primarily because of the rarity of preservation. Ethnographic accounts of the re-use of roof timbers from old houses in semi-arid regions are one example of the likely considerable curation and reuse of wood in timber-poor areas. Even in timber-rich areas, the effort needed to create planks or carved wooden architectural pieces led to considerable curation and reuse.

The investigation of these categories of extractive-reductive crafts—stone, fiber, and hard organic materials—has already revealed similarities and differences in the techniques of production, the organization of production, and consumption patterns. In the next chapter, I examine three groups of pyrotechnologically transformative crafts from a similar comparative perspective.

CHAPTER 4

Transformative Crafts

Most illustrious Princes, often have I considered the metallic arts as a whole, . . . just as if I had been considering the whole of the human body; and when I had perceived the various parts of the subject, like so many members of the body, I became afraid that I might die before I should understand its full extent, much less before I could immortalize it in writing.

(Agricola 1950 [1556]: xxv)

In this chapter I again provide overviews of the basic production processes for three groups of crafts. This is done to establish a basis for understanding the practice of these crafts and the role of their products and practitioners in societies. It will also allow the comparison of craft industries, as discussed in Chapter 7. In this chapter, the craft groups are all transformative; that is, the transforming of the chemical or micro-structural properties of their raw materials is an essential part of their production. Soft, pliable clay, readily dissolving in water, is transformed into hard, relatively impermeable terra-cotta, stoneware, or porcelain. Rocks and minerals are transformed into metals that can be shaped into almost anything by hammering or by casting, then re-melted and reshaped again. Finally, perhaps the most amazing transformation of all, opaque, everyday quartz pebbles or sand are transformed into translucent, extraordinary glass. All of the craft groups discussed here are pyrotechnologies, transforming materials with the use of fire (or more accurately, heat). Chemically transformative crafts were relatively rare in antiquity, other than dyeing, and not all dyes functioned by chemical transformation as described in the Fiber section in Chapter 3. The production of lime and gypsum plasters and mortars, as well as cement, are excellent examples of both pyrotechnologically and chemically transformed materials (Hodges 1989 [1976]), and the two former were some of the earliest transformative technologies in the world.

The first two groups of crafts covered in this chapter, fired clay and vitreous silicates, are ceramic materials; they are fine-grained materials that can be shaped in an additive rather than a reductive fashion, and which are hardened by heating. Such a definition of "ceramic" follows standard materials science usage, and is not interchangeable with either "terracotta" or "pottery," contrary to common usage in everyday language and in much of the archaeological literature. Fired clay was used to create many objects, but one type, pottery (fired clay vessels), is a favorite artifact of archaeologists. With its high degree of preservation, and its ability to be the carrier of technological and symbolic information through fabric, form, and decoration, it has been used for a staggering number of insights into all manner of archaeological questions, not the least of which is basic chronology. The vitreous silicates, here represented by the overlapping categories of glazes, faiences and glass, are all ceramic materials as well but are formed primarily of particles of quartz. Unlike the other craft groups discussed in this book, all of these vitreous materials were developed only across Eurasia, and were not produced in the Americas prior to European contact (Rice 1987: 20). It is still not clear how these vitreous silicates are related, whether glass evolved from experimentation with faiences or glazes for example, although there are strong reasons to support such a link (Henderson 2000: 54). Nor are the relationships clear between the development of these crafts in different regions, although there are known differences in some of the varieties of faiences found in different parts of Eurasia. There may have been a single center of invention with subsequent diffusion of knowledge, or completely independent invention of these materials in different regions. Personally, I think the most likely scenario was something like the historic case for porcelain, albeit on a less complex scale. Some region may have been the first to develop the manufacture of objects made from these new materials, but independent invention using slightly different techniques occurred rapidly in other regions, triggered by attempts at copying traded objects. With increasing availability of well-analyzed, solidly dated material from the different regions, the technological history of these vitreous silicate materials should start to become clear. Finally, the Metals section focuses on copper and iron, the two major metals of the pre-modern period. The fascinating diversity of production methods for these metals and their alloys, not to mention other metals such as gold, silver, and lead, are not given the space they deserve—as is true of every other aspect of technology covered in this book. Instead, I have created a basic outline for comparison with other crafts, and reference some of the excellent summary volumes of metal working for further introduction to this field. Fortunately, metal technologies continue to be the focus of considerable archaeological, experimental, and ethnographic study in numerous areas of the world, so there is a great deal of informative literature available.

In Chapter 4, I focus almost entirely on production processes for these crafts, although I do incorporate descriptions of the social and economic settings in which craftspeople might make production choices. As in Chapter 3, my focus is on production rather than analysis, so where a difference exists I have described materials and processes from the perspective of the producer rather than the analyst. Unlike Chapter 3, however, in this chapter I do not separately examine selected aspects of organization of production and consumption for these three craft groups. Instead, there are three topical studies in Chapters 5 and 6 specifically focused on organization of production, social aspects of consumption, and cultural aspects of production choices for these three technology groups. The section on Labor in Chapter 5 includes a focus on pottery production as an example of craft specialization, my discussion of Technological Style in Chapter 5 examines metal technologies in various parts of the world, and the section on Value and Status in Chapter 6 uses vitreous silicates as a primary example. In all three cases, I look at the technological system as a whole, as well as how these systems fit within particular economic, political, or social settings.

FIRED CLAY

> ... their material equipment, their huts, their fields, in fact everything around them is composed of clay.
> (Shah 1985: 148)

As noted above, fired clay encompasses one of archaeology's most important categories of artifacts, pottery or fired clay vessels. Bricks and other building materials—tiles, drain pipes, and decorative pieces—are created from fired clay as well. An enormous number of other sorts of objects are also made from fired clay, the most common worldwide being figurines, but also beads, spindle whorls, toys, braziers, and so forth. Rice (1987) is still the essential archaeological reference on pottery production, consumption, and analysis, and she has updated her summaries of pottery analyses on specific topics in two more recent papers (Rice 1996a, 1996b). Shepard (1976) is still an important resource for analysis; Rye (1981) is a significant source for ethnographic production and archaeological identification; and Sinopoli (1991) provides excellent case studies on central archaeological topics addressed through pottery analysis. Orton *et al.* (1993) provide good grounding in typological analysis, particularly from a paste-based approach. Henderson (2000) covers more recent methods of analysis than Rice, and provides case studies on high-fired pottery and glazed ware production. Each region of the world also has its specialist works on pottery analysis as well as other clay materials; Dales and Kenoyer (1986) is such an example for my area of specialty, the Indus civilization.

Fired clay objects were made from different types of clay, fired to different ranges of temperatures. I loosely use the term clay to refer to "fine-grained earthy material that becomes plastic or malleable when moistened," as Rice (1987: 36) so succinctly puts it, primarily minerals derived from high-alumina silicate rock (Al_2O_3 and SiO_2), particularly feldspar and mica. Rice (1987) provides extensive details about clay types and temperature ranges, including a comprehensive discussion of clay from a geological, mineralogical, chemical and potters' viewpoint (see also Henderson 2000; Rye 1981). Analysts typically describe types of clays using geological terminologies, and group clay bodies in this way; Table 2.7 in Rice outlines the properties of the four major groups of clay minerals. The composition of the *clay body* (the clay plus any added materials) has a large impact on the properties of the objects created from it. Archaeologists also use composition, as determined by petrography and chemical/physical analyses, to determine the origins of the materials employed, and hence (roughly) the origins of the fired clay object. For potters and other workers in clay, however, it is the desired properties of the final object and the working properties of the total clay body that are the most important points. A crafter of clay must achieve, through a variety of methods, a clay body and a final product with the working and final properties desired.

I will employ two broad groupings for fired clay materials, from Hodges' (1989 [1976]) division of fired clay bodies into those fired below the sinter point (primarily iron-rich terracottas and earthenwares) and those fired above the sinter point (stonewares and porcelains) (Figure 4.1). This is not to denigrate the importance of identifying the clay body from an analytical viewpoint as well, and Rice (1987) explains in detail the chemical and physical processes occurring in the production of each of these fired clay groupings. Hodges (1989 [1976]) defines the *sinter point* as the temperature range within which clay particles fuse together (sinter), producing a dense, impervious clay body. The sintering point of clays can be lowered by adding various materials that act as *fluxes* (materials which lower the melting point), such as mica, potash, lime, or bone, depending on the clays. Rice (1987: 94) notes also that ceramics made from finer particles sinter and vitrify at lower temperatures than coarser textured clays, regardless of the type of material, a point of special interest in the fritting of quartz for faience, glaze, and glass production as well. Rye (1981: 106) comments that sintering is not only affected by the absolute heat of firing, but also by the length of time at which the heat is

Fired Clay Body Type	Sintering & Porosity	Firing Range	Typical Applications	Glazing?, Color, etc.
Terracotta	Unsintered; High porosity	Below 1000°C	Most prehistoric pottery; bricks, roof tiles, flowerpots	Mostly unglazed, but some glazed; Often iron-rich, red-fired
Earthenware	Unsintered; Medium porosity	900–1200°C	*Fine*: majolicas, wall tiles, Roman Arretine/Samian ware; *Coarse*: bricks, drainpipes, tiles	Glazed (especially finewares) or unglazed
Stoneware	Sintered; Low porosity	ca. 1200–1350°C	Tableware, decorative objects, glazed tiles & drainpipes	Glazed or unglazed; Often buff-brown color; Fractures like knapped stone
Porcelain	Sintered; Very low porosity	1300–1450°C	Fine tableware, decorative objects, modern equipment	Translucent, hard, fine; White; 'Rings' when tapped

FIGURE 4.1 Fired clay types (adapted from Rice 1987: Table 1.2 and Hodges 1989[1976]: 53).

FIGURE 4.2 Vitrified pottery, forming a pottery "waster."

applied; as temperatures increase, the time required to achieve permanent sintering decreases. If the clay body is heated too far above the sinter point, the clay particles melt and *vitrify*, causing the object to collapse (Figure 4.2). Many, perhaps most, archaeologists use the term "vitrify" more like my use of the term "sinter," as Rice does (e.g., Rice 1987: 5–6). I prefer to reserve the term "vitrify" for collapsed, over-fired, failed clay objects, as does Hodges, Rye, and others, as I need this separate term to clearly distinguish the many melted, discarded, fired clay fragments (slags) that I examine from several different pyrotechnologies.

Why is firing above or below the sinter point important? The primary reason is the degree of control of the firing necessary to fire above or near the sinter point; the potter had to be very knowledgeable and have well-developed, reliable kilns to repeatedly fire near the sinter point without ruining a kiln full of pottery by vitrifying it. Understandably, objects in the second group, made from clays fired above the sinter point, were often display items used to connote status, and wealth. These materials were developed much later than the lower-fired clays in every region where they were found. There would be very visible differences between the two types of fired clay objects, those fired above the sinter point being very dense and impermeable, and very hard and brittle. In some cases, the latter were even translucent.

As shown in Figure 4.1, *terracotta* objects are low-fired (ca. 500–900°C), and usually made from iron-rich clays fired red under oxidizing conditions. Terracottas are the most ubiquitous material used in the past for pottery, figurines, bricks, and other objects. *Earthenwares* are made from the same or similar sorts of clays, but fired at higher temperatures (ca. 900–1200°C). The materials fired above the sinter point, typically above ca. 1000°C, are stoneware and porcelain, which were only produced in Eurasia. True *stonewares* are usually made from iron-poor buff or light brown clays, and get their name from the stone-like way in which they fracture, like chert. The highest fired clays, *porcelains*, are made from white kaolin clays to create a translucent, very hard product. This is a very simplified terminology, and there are plenty of exceptions and difficult to characterize fired clay bodies. For example, "true" stonewares are found starting in the second millennium BCE. However, earlier materials from the third millennium BCE are also called "stoneware" because they fracture like chert when broken, and because they have extremely low porosity (Schneider 1987). These materials are known from both Mesopotamia, used only to make a specific type of pottery (Stein and Blackman 1993), and from the Indus Valley, used only to make bangles (Vidale 1990; Blackman and Vidale 1992). However, the third millennium stonewares were fired to lower temperatures (between 1000 and 1100°C) and were made from the iron-rich clays used to make terracottas, but fired in a reducing atmosphere to achieve deep black colors. Nevertheless, because they are sintered clays, they fall into the category of stoneware rather than earthenware. The categories of "chinas" are even more complex. China is sometimes used as synonymous with porcelain. Others distinguish chinas as white clay bodies containing a variety of fluxes and sometimes other clays, made in imitation of porcelains; many of these chinas are also translucent, but are not as hard as porcelain (Rice 1987).

These clay bodies could be painted, incised, and/or glazed, to create different decorative surfaces. The study of pottery decorations is the most prevalent type of stylistic study. Pottery decoration has been one of the major datasets for archaeologists looking for information on group identity, reinterpreted over and over again as different methodological and theoretical approaches waxed and waned in popularity. Potters have been one of the most common ethnoarchaeological objects of study. Pottery, both its surface and fabric, has informed on status, on trading networks, and on learning methods. Pottery analysis has been the source of several major approaches to technological systems, most notably that of ceramic ecology. My general outline of fired clay object production thus privileges the processes associated with pottery manufacture.

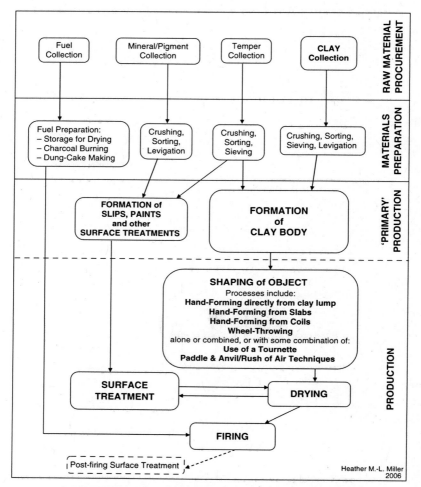

PRODUCTION PROCESS DIAGRAM FOR FIRED CLAY (pottery)

FIGURE 4.3 Generalized production process diagram for fired clay, focused on pottery (greatly simplified).

The general production of fired clay objects employs the following stages (Figure 4.3):

1. Collection of clay, temper materials, any needed decorative pigments, and fuels
2. Preliminary processing of clay (cleaning, sifting, soaking, levigation); preparation of temper (crushing, cutting, sieving); preparation of decorative pigments and other materials

Transformative Crafts

3. Formation of the clay body (mixing, kneading, maturing)
4. Shaping/fabrication of clay objects, employing (a) hand-forming, (b) molding, (c) use of turning devices, (d) trimming/scraping, and/or (e) paddle and anvil
5. Drying of objects and surface treatments (painting and/or slipping, incising or impressing, polishing, smoothing)
6. Firing of objects.

Most objects are finished at this stage; some have additional steps (not shown on Figure 4.3):

7. Further surface treatments ('seasoning', post-firing painting, applying glazes to biscuit-fired ware)
8. Second firing of object (for glazed wares and porcelains).

COLLECTION AND PRELIMINARY PROCESSING; FORMATION OF THE CLAY BODY

The availability of raw materials is a major concern for craftspeople, and the supply of raw materials is often suggested as a method of control of production. This is seldom a major issue for terracotta production, although clay deposits were sometimes owned and controlled. In large riverine floodplains around the world, clay suitable for terracotta production was generally widely available. Particularly well-sorted pockets of clay and/or sand exist, so that modern-day potters do identify preferred locations for the collection of clay, but on the whole clay sources are quite abundant. In other geological landscapes where there is more diversity of sediments, suitable clays may be more difficult to find; glacial landscapes have particularly unpredictable pockets of clays. Although *ideal* clays may not be available, there are few parts of the world without clays capable of at least coarse terracotta production. Clays suitable for more specialized or higher-firing products are a great deal more rare, so that access to clays might be more difficult, and would need to be considered when investigating past production. Ceramic ecologists, notably Dean E. Arnold (1985), have examined the question of raw material procurement in great detail through ethnographic studies of potters.

Clays are collected by digging the sediments out of the ground (Figure 4.4a) and transported back to the potter's workshop. Even the "best" clay then requires some processing for vessel and object production, particularly for wheel-throwing of vessels. It is spread out and beaten to break up the clods, and any roots or inclusions are picked out by hand (Figure 4.4b). Ethnographically, apprentices or helpers often performed this stage of clay preparation as well as the potters themselves. If necessary, the clay is passed through a

FIGURE 4.4 (a) Collecting clay and (b) processing dry clay by sorting out unwanted inclusions and breaking down lumps.

basketry, cloth, or metal sieve of the desired fineness to remove any remaining inclusions. If very fine clay is desired, the clay may also be levigated—placed in water and agitated, then allowed to settle out into size classes. Levigation can be done in a vessel and decanted in layers, perhaps through a sieve. It can also be done in a pit, lined with clay to prevent sediment inclusions, where the water is allowed to evaporate and the clay then removed in layers, with the finest clays at the top as these are the last to settle out of suspension. Once the clay is processed to the desired quality, it is mixed with water (*slaked*) and left for a day or two. Any desired tempering material is added, and the mass of clay is kneaded with hands or feet to work the water and temper thoroughly into the clay, producing a uniform clay body. This prepared clay may be wrapped to keep it damp until the potter is ready to use it, or placed in a covered basin or pit. The clay or clay body can be stored for considerable periods of time at various stages throughout this preliminary processing, when unsorted, sorted, mixed with temper, or slaked and kneaded. Sometimes storage of the final clay body prior to use is said to improve its working qualities.

Temper refers to any type of material added to the clay. Since we cannot always tell from the resulting products whether materials were added to the clay or were found in it naturally, some archaeologists prefer the term "inclusion" rather than "temper." *Inclusion* simply refers to the presence of non-clay materials in the clay body, with no suggestion of whether these materials are found naturally in the clay or deliberately added. Rice (1987: 406–413) provides a more extensive discussion of terminology relating to this point. Inclusions, whether deliberately added or not, affect the working and firing properties of the clay. Some clays did not need any added tempers, but functioned as desired without any additional materials beyond whatever was found naturally. Other clays, or other purposes, required the addition of tempering materials to achieve the desired working or firing properties. Special materials might also be added for ritual reasons. Common inclusions or tempers are plant materials, sand, shell, mica or other minerals, grog (fragments of fired pottery or brick), dung, salt, or other clays. These materials generally do not require much processing prior to mixing with the clay. Plant materials can include straw or seeds from domesticated or wild grasses, seed fluff, leaves, and so forth; these need only to be chopped to the required size. Adding dung can provide pre-chopped plant materials of this kind, as well as other organic materials. Plant materials typically burn out during firing, leaving voids and sometimes silica skeletons (phytoliths) behind, which add to the heat resistance of the final product. Sand, shell, mica and other minerals, and grog may be ground or sieved to procure the desired size of particles. Addition of salt can counteract some of the negative properties associated with shell or other calcareous inclusions, as Rye (1981) has elegantly demonstrated. Acquiring desired tempering materials sometimes required greater effort than acquisition of the clay itself, and such materials might come from farther away.

Materials used for surface treatments include mineral pigments, clays, and sand for the production of slips and pigments, and tools for incising, impressing, or stamping patterns. The production of glazes would also require fluxes such as plant ash or minerals, as discussed in the next section on vitreous silicates. These surface treatment techniques are defined and discussed in detail in the Surface Treatment section below. On the whole, few of the materials used for surface treatments required elaborate processing prior to their use, other than glazes. The minerals used for slips and pigments were sometimes the most difficult materials to acquire. These materials might not be available in the local landscape, but had to be acquired from considerable distances, as ceramic ecologists have documented. The minerals used for pigments were thus often the greatest expense for potters, except perhaps for fuel. Coloring minerals such as red ochre or manganese-iron compounds would have been crushed and/or ground to a powder, and mixed with fine clay and water to the desired consistency. Alternatively, the minerals might have been soaked in water before and/or during crushing or grinding, to lubricate this process and perhaps to soften the mineral. Less water would then need to be added to the paste created. Sand, grog, rock minerals, or other materials added to slips were often ground and sieved to select the size range desired. Materials used to make glazed surfaces could be very diverse, sometimes requiring extensive trade networks and precise knowledge for their procurement and use. Most of the tools used for creating incised or impressed designs were relatively simple, from easily available materials, but a few such tools required elaborate manufacturing of their own, particularly the molds used to create surface patterns, discussed below. For glazed object firing, the potter would usually need to make firing containers and setters to avoid contact with fuels and smoke, if a single-chamber kiln was employed, and to help ensure proper stacking and avoid sticking of the glazed objects.

Fuel is a major raw material required in quantity by all of the high-temperature pyrotechnologies, so that its supply was an important issue for craftspeople. Wood fuel was used in most terracotta firings, and wood from specific species might be selected for desired characteristics of heat or smoke production if opportunity afforded. Prepared wood in the form of charcoal might be used for higher temperature fired clays, or for those firings requiring relatively smoke-free firing (such as glazed wares), but at added expense. There are both ethnographic accounts and some archaeological evidence that "waste" fuels were used for firing terracotta objects, especially in more ephemeral firing structures with less strict atmospheric control. Waste fuels are agricultural by-products such as chaff, straw, tail grain, and oil pressings, as well as gleaned twigs and small sticks; they are a much less costly alternative to wood or charcoal. Finally, some waste fuels and especially dung fuels were deliberately chosen for the creation of black-fired objects, either for their high organic content where reducing firings were desired or for their high smoke production where sooting was employed. (See below in the Firing section for definitions of these terms.)

The fuels used in ancient firings can be determined through archaeobotanical analysis of material from excavated firing structures of different types (Goldstein and Shimada in press). By showing a contrast to fuels found in other contexts, such as cooking hearths, this research is able to determine if specific types of woods and/or other materials were deliberately chosen. To investigate the possibility of more than one type of fuel being used, bags of samples must be collected to allow analysis for twigs, dung, and agricultural waste, rather than simply selecting large pieces of charcoal. As it is likely that different kinds of fuels were used in different kinds of firing structures, it is important to collect such samples from more ephemeral structures as well as from updraft kilns.

Shaping Methods

In the limited space available here, I can only sketch a rough picture of the most common shaping or fabrication techniques for the great variety of fired clay objects. Much more extensive and specialized discussions can be found in the articles and books listed at the beginning of this fired clay section. Most of my definitions are after Shepard (1976), Rice (1987), or Rye (1981). Figure 4.3 provides an overview of the place of these various fabrication techniques within the overall process of fired clay object production, primarily for terracotta vessel production.

Fired clay objects are hand formed, formed in molds, or formed with the use of turning devices. The shapes of objects created in these ways can be altered with the use of turning devices (as when a coil-built pot is further shaped on a turnable support), by trimming and scraping, or by the use of a paddle and anvil. Common *hand-forming* techniques are pinching or modeling, slab formation, and coiling. Pinching and forming or modeling a ball of clay by hand has been used to produce vessels, figurines, ornaments, and many other objects. It is useful for small objects, unique objects, or for clays with difficult working properties. Except for these cases, it is seldom used for production on a large scale, but is not uncommon for the small-scale production of vessels. Production and assembly from *slabs* can also be used for pottery construction, but has most often been employed for the production of architectural objects (e.g., roof and floor tiles) and other non-vessel clay objects. The other major hand-building technique, *coiling*, has been an extremely common method of vessel production worldwide. Coiling has also been used to make other small terracotta objects, as well as large sculptural pieces such as the legs and torsos of the famed Chinese terracotta soldiers. Production of vessels using coiling is accomplished by manufacturing cylinders or ropes of clay, winding these on top of each other in a circle to build up the vessel walls, then smoothing the joins between the coils until they are no longer visible. However, the coils can often still be distinguished by xeroradiography, and the use of coiling versus wheel-throwing can also sometimes be distinguished by various visual characteristics, including breakage patterns and the distribution and orientation of inclusions in an assemblage.

Molding was used to form complexly shaped vessels, such as the elaborate portrait vessels of the Moche of South America (Donnan 2004). Molds can also be used to form and support portions of large vessels, such as the bases of large jars, until the clay dries and can support its own weight without slumping. Molding allowed mass production of figurines, as seen throughout the Hellenic world, as well as small angular vessels such as Roman lamps. Molds were also used to mass-produce vessels with elaborate impressed or raised designs like the Arrentine or Samian wares of the Roman empire and the historic water vessels of South Asia. Usually, clay was patted into or over an appropriately shaped mold (perhaps after first dusting the mold with fine ash, sand, or some other "parting" material), and shaped to the mold's surface. Potters often made such molds themselves out of fired clay. Another molding method, slip-casting by pouring liquid clay into molds, appears to be a relatively recent phenomenon (Rice 1987).

The term *wheel-throwing* is often casually used to cover a range of forming techniques using a range of *turning tools*. However, it is important to distinguish between the use of "true" wheels, the use of tournettes, and the use of turnable supports (Figure 4.5). All three of these classes of tools allow vessels

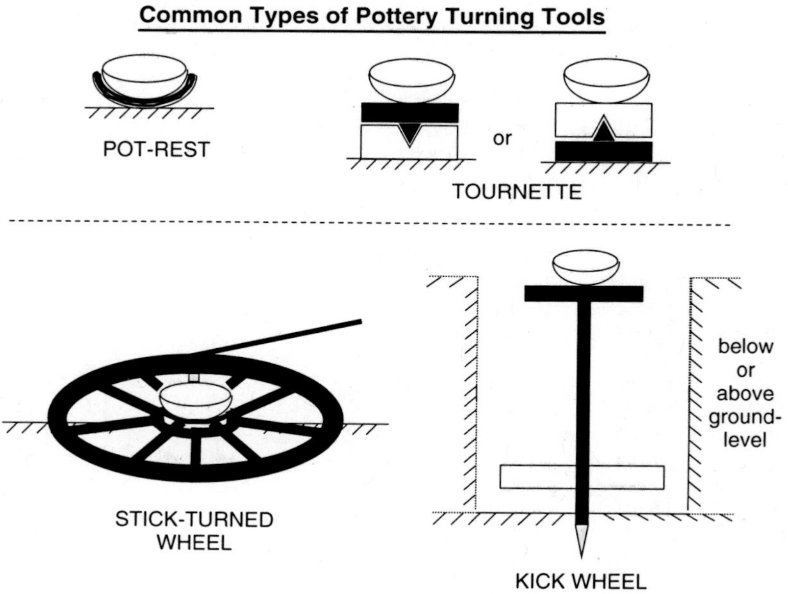

Common Types of Pottery Turning Tools

POT-REST

TOURNETTE or

STICK-TURNED WHEEL

KICK WHEEL

below or above ground-level

FIGURE 4.5 Types of turning tools for shaping and decorating clay vessels.

Transformative Crafts

FIGURE 4.6 Zaman evening the base of a large vessel on a kick-wheel. The same kick-wheel is used as a tournette to add coils to the upper surface of large vessels whose bases are made in molds.

in the process of formation to be rotated so that the potter need not move around the vessel. All of these tools can be used for any stage of production, from initial forming through modification, finishing, and surface treatments (Figures 4.6 and 4.7). However, "wheel-throwing" is technically only applied to the use of true wheels rotating at a rapid speed for a considerable period of time (Rice 1987: 132–134). A *pot rest* or turnable support is any object on which clay can be placed and formed with one hand while the other hand (or the feet) rotate the support (Figure 4.5). They are often made from the bases of fired round-bottomed vessels, but the term can apply to any turnable support without a pivot. *Hand-wheels* or *tournettes* (from the French *tourner*, "to turn") not only allow rotation, but also have a pivot to center the revolutions (Figure 4.5). Like fast or "true" wheels, tournettes are spun by the hands or feet of the potter or an apprentice. Since tournettes (sometimes mislabeled "slow-wheels") have a pivot, they can be rapidly rotated for a short period of time, and may even produce surface marks characteristic of wheel-thrown pottery, such as rilling. (See Rice and especially Rye for discussions and photographs of surface marks indicative of various methods of pottery production.) However, only "true" wheels, which produce centrifugal force as well as sustained momentum about a central pivot, can consistently produce characteristically wheel-thrown pottery. Centrifugal force and momentum are produced and maintained because these wheels can sustain rotation at high

FIGURE 4.7 Nawaz painting the upper surface of a water jar on a kick-wheel used as a tournette.

speeds for considerable lengths of time in spite of friction produced by the potter manipulating the clay. This is done either by the use of a flywheel, as in kick-wheels, or by sheer weight, as in spun- or stick-wheels (Figure 4.5). The *kick-wheel* is a double wheel, consisting of an upper wheel upon which the vessels are formed connected by a shaft to a large, heavy lower wheel ("flywheel") which is turned by the potter's foot. The shaft is connected rigidly to both wheels, and continues through the other side of the lower wheel to be embedded into the ground or floor, forming the pivot around which the entire structure turns. The kick-wheel is thus a rather complex, permanently located piece of equipment, but can be continuously used. The spun-wheel or *stick-wheel* is a heavy single wheel set directly on a pivot, and usually raised only a decimeter or so off of the ground. Such wheels can be solid or spoked around a central turntable. They are spun to a high speed using a pole inserted into a slot at the outer edge of the upper surface. The potter then removes the pole and works the clay until the heavy wheel has "wound down," rather like a toy top. The process is then repeated, as it is not possible to turn the wheel with the pole while the potter is at work. Wheel-throwing is usually a much faster production technique than coiling, in spite of the need for special equipment, as fast wheel-throwing required less time per vessel produced. Fast wheel throwing allows very rapid production of vessels, and is primarily associated with mass production by specialized potters. Other types of turning devices are widely used in slower, smaller-scale production such as the household level, but are also used by specialized producers, so care must be taken in using wheel-throwing as a proxy for production scale. The lack of fast wheels altogether in the Americas is a case in point. The use of the fast wheel for mass production is further discussed in Chapter 5, in the section on Innovation and Organization of Labor.

Many vessels and other objects are further shaped either while still wet or after drying to the leather-hard state. These additional shaping stages are sometimes referred to as "secondary shaping" but I have grouped all shaping techniques together, in part because the same techniques can be both "primary" and "secondary," such as the use of turning tools. Various turning tools could be used at various speeds to produce these shape modifications, such as altering or evening the shape of a still wet coil-made vessel (Figure 4.6). Common modifications are *scraping and trimming*, particularly of vessel bases, with stone, bone or fired clay tools. This is frequently but not exclusively done with the vessel placed upside down on a turning device. Whether such scraping and trimming was done before or after the vessels were leather-hard can be determined by characteristic marks on the vessels (Rye 1981), and/or by usewear marks on the associated cutting tools (e.g., Anderson-Gerfaud, et al. 1989; Méry 1994). Another common shape modification technique used by potters around the world employs a *paddle and anvil* to smooth, re-shape and thin walls and bases of coiled or wheel-thrown vessels, usually to produce rounded bases. Such paddle-and-anvil forming is done on slightly dry but not leather-hard vessels produced by one of the methods described above. Modern potters in South Asia use a fast technique requiring no tools which produces results similar to paddle-and-anvil techniques (du Bois 1972). After the vessel has been formed, it is turned upside down and the mouth is rapidly and forcefully slapped against a flat surface. The inrush of air puffs out the base to form a well-rounded bottom. Only certain types of clays and very even-walled construction allow the use of such a technique. Finally, pieces can be *applied or joined* together, such as the application of handles onto a vessel or the joining of body parts on a small figurine or a large sculpture. As is apparent from this discussion, combination of techniques are frequently used to produce vessels, so that a single vessel might have a mold-made base joined to a wheel-thrown (and perhaps previously coil- or slab-built) body and rim. The base might then be further modified by using a paddle and anvil to round out the bottom. On the other hand, many of the smaller and mass-produced vessels were entirely thrown "off the hump" using a fast wheel. It is difficult enough to determine the variety of shaping methods ethnographically; when confronted with a finished archaeological product, the few remaining clues require patient and detailed observation of a large number of vessels to untangle the process of production.

Drying and Surface Treatments

The need for a place to dry and store fragile but frequently large and bulky unfired clay vessels and other objects is often overlooked in discussions of pottery production. Immediately after forming, the objects need to dry to a leather-hard state. This can be a lengthy process during wet and cold times of the year, which is one reason why pottery making is often a seasonal activity. Clay objects require shelter from both moisture and from extreme heat, as drying too rapidly can lead to cracking, particularly for relatively fine clays. These objects, still fragile, must be stored until various forming and surface treatments have been performed, from scraping to painting, and again after such treatment until enough vessels have been produced to fill a firing structure. This is a serious space requirement in crowded city contexts. Even in smaller communities with more space available, some sort of storage area would need to be constructed, at least a roofed shelter. The third millennium BCE site of Naushuro in Pakistani Baluchistan provides a very rare example of such a storage space, complete with the remains of the unfired clay vessels originally placed on shelving while awaiting slipping and painting, and even trimming scraps (Méry 1994). For large vessels, particularly those built in parts (such as molded bases with wheel-thrown bodies/rims), drying includes an additional issue besides storage space. Moving large vessels off the wheel while still relatively wet could result in disaster. However, letting the vessel dry on the wheel monopolizes the wheel, restricting production. Perhaps in special cases, care was taken to make such vessels at the end of the day, allowing overnight drying on the wheel. However, at least in modern times, the use of a bat helps tremendously with the removal of large and/or especially fragile vessels from a wheel. A *bat* is a thick disc of wood or fired clay attached in some fashion to the wheel (whether by clay or by pegs and slots), which rotates as part of the wheel during its use. The clay to be thrown is placed on the bat, the vessel is formed, then the bat as well as the vessel is removed from the wheel. This provides a rigid base to support the vessel and provide hand-holds during removal and transport of the wet vessel. Heat must sometimes be provided for drying stages, although it creates an extra fuel expense. In temperate climates, when even the hot season might be wet, objects might be dried in a low-temperature oven. Drying ovens were also used in the production of higher valued or elaborate products, such as the production of glazed wares and porcelains.

Surface treatments for fired clay objects include the application of slips, pigments, or other materials; incising or impressing, or otherwise moving the clay surface; and polishing or smoothing, which could be included with "moving" the clay surface. Some of these treatments are done prior to drying, but most are done after the object has dried to some degree. A few types of pigments are applied after firing, but most treatments are usually done before firing. *Smoothing* of a surface is done with a soft tool such as a cloth or the hands, usually when either the clay or the soft tool is wet, to even out the surface and create a fine finish. *Polishing* is done with a hard object such as a pebble or piece of wood when the clay is leather-hard, and creates a glossy surface by orienting and compacting the clay particles on the surface of the object. Fired clay objects can be *incised or impressed* with various tools including complex stamps, grooved with fingers, combed in wavy lines, impressed with cordage, or incised with symbols in complex design patterns. "Wet wares" are created by impressing cloth, matting, or other patterned objects into the wet surface of a vessel. Creating impressions on a vessel with a cord-wrapped paddle is a well-known surface treatment found throughout the Great Lakes region of North America. Incising and impressing can be done either before or after the vessels are leather-hard, and occasionally after firing. Rice and Rye provide many more examples of impressed and incised surface treatments, as do the many region-specific works on pottery. Incised and impressed patterning is probably the most widespread method of decoration worldwide through time, although there are definite shifts in the relative importance of incising/impressing and slipping/painting in different time periods and places. Groups using very elaborate slipped and painted decorative designs seldom made extensive use of incised or impressed designs, and vice versa. Of course, there were exceptions and some combinations did occur, for example the use of grooved or molded designs together with slipping, or outlining the edges of painted designs by incised lines.

The application of slips, pigments and other materials includes slipping, painting, and glazing, primarily to apply color but in some cases to attach small particles of other materials such as sand. The terms slip and pigment are clearly defined by Rice (1987: 148–149) and Rye (1981: 40–41). *Slips* are fluid suspensions of fine clay and other coloring materials (primarily minerals) applied over a sizable proportion of the object, usually prior to firing. For example, Rice (1987: 49) notes that illites are excellent clays for making slips, as they have a natural luster. *Pigments* are also fluid suspensions of fine clay and other coloring materials, but are usually slightly more viscous and are applied with a brush to form designs on an object prior to firing.

There is clearly great potential for overlap and confusion between these terms. Pigments are also sometimes called 'paints', but both Rice and Rye emphasize that the term *paint* should be limited to the action that is used to apply a substance, not the nature of the substance used. The vast majority of slips and pigments are colored by various forms of iron, which fire to different colors depending on the atmospheric conditions of the kiln. Under oxidizing conditions (that is, where oxygen is plentiful in the kiln during firing), most iron-based slips and pigments fire red or orange. Pigments or slips containing manganese iron compounds will fire black or brown or purple under oxidizing conditions. Other mineral and atmospheric combinations result in different colors. It is not always easy to differentiate "true slips" from "self-slips." A true slip is produced by the application of a colored material to the surface of a clay object, usually prior to firing but sometimes afterwards. (Note that a slip applied after firing is sometimes called a wash (Rice 1987), although this term is used in a variety of ways.) Pottery can also have a *self-slip*, a surface coloration derived from the smoothing action of the potter's hands or a cloth at the end of the manufacturing process, bringing a layer of fine clay particles to the surface (Rice 1987), perhaps combined with particular kiln atmospheres. A self-slip can also be due to the presence of soluble salts in the clay, which migrate to the surface of the vessel as it dries, leaving a white, cream or yellowish coating. Such self-slips can thus be either intentional or unintentional (Dales and Kenoyer 1986: 63; Kenoyer 1994: 358–359). Slips can also contain larger particles; for example, sandy slips are often rough, thickly-applied exterior slips containing sand. Grog (ground up pottery sherds or bricks) and rock minerals are also applied in slips. Slips can contain both colorants and larger particles, probably simply by adding these particles to the usual colored slips. These coatings can have functional purposes; for example, sandy and other large-particle slips were and are applied to the lower bodies of cooking vessels, with the sandy coating providing resistance to thermal shock from cooking fires (Rice 1987: 232; Schiffer, et al. 1994). Of course, some of these coatings may have been primarily decorative as well, and some were certainly incorporated into design themes by incising patterns into them. Colored slips most often covered the visible surfaces of an object as decoration, but colored slips are also sometimes found on the interior of terracotta jars or bottles for functional reasons, to reduce the porosity of the vessels and allow them to be used for the storage of liquids.

Both colored and other types of slips can be applied by immersing the object in the slip, by pouring the slip over the object, or by applying the slip over the surface with a cloth or sponge-like substance. *Painting* is the application of pigments to form designs on a surface, usually with a brush or other thin object. Painting could be done over slips or directly onto the clay surface, while holding the object in the hand or while it is on a turning device (Figure 4.7). Tournettes and turnable supports are particularly well suited for painting, as neither rapid nor continuous speed is necessary, and pots with rounded bases can even be spun on their own bases. The application of designs is usually done free-hand, although for elaborated painted designs, the outlines of the figures or design boundaries are sometimes faintly laid out using measuring devices or compasses. Polychromes were slipped and/or painted with three or more colors, and resists of wax or other materials have been used in some societies for very elaborate designs. The production of glazed objects and porcelains include very elaborate sequences of surface treatments, some of which are referenced in the section after firing.

Firing

The next step is to load the vessels and/or terracotta objects into the firing structure. Methods of loading are dependent on the type of structure used and on the assemblage of objects to be fired. The placement of objects in the firing structure is an extremely important stage of the process, as incorrect placement can result in poor firing, marring of surfaces, or even the destruction of the products. Some of the firing tools or 'kiln furniture' such as setters, saggars, or other containers were designed especially to prevent marring by separating stacked vessels and/or enclosing objects to avoid contact with fuel or smoke. Particularly in multi-chamber kilns, where fire-clouding is greatly reduced by the separation of fuels and objects, a very even surface can be achieved if the pieces are correctly loaded and fired. On the other hand, especially in ephemeral and single-chamber firing structures where the objects are in direct contact with the collapsing fuels, the shifting of poorly loaded vessels during the firing can result in the destruction of a large portion of the products.

Pre-industrial firing structures for clay can be divided into three basic types: ephemeral firing structures, single-chamber firing structures, and multi-chamber firing structures. These categories are based primarily on nature of the draft (heated air flow) system, and to some extent on atmosphere control mechanisms. Note that these categories do not necessarily represent a progression in temperatures reached; ephemeral firing structures and simple pit single-chamber structures have been measured at surprisingly high temperatures, particularly with certain types of fuel. However, these three categories do have differing degrees of control of the sustainability and evenness of temperature production. Basically, *ephemeral firing structures* include "bonfire" structures, where vessels are fired in the midst of an open fire and just surrounded by fuel (e.g., May and Tuckson 1982; Rhodes 1968; Rice 1987: 155), as well as some types of fairly substantial covering in addition to fuel, including potsherds, earth, and/or mud plaster (Figure 4.8a) (also

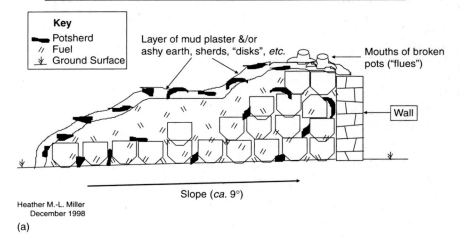

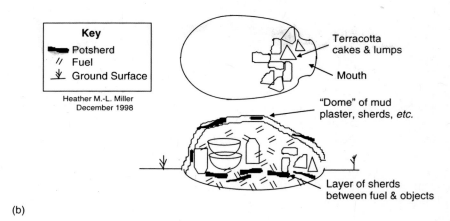

FIGURE 4.8 Various firing structure types for firing pottery: (a) example of one type of ephemeral firing structure, an "open-air" firing structure; (b) example of one type of single-chamber firing structure, a pit kiln. (See Miller 1997 for full descriptions of these particular structures.)

see Sinopoli 1991: 39). However, these structures are neither dug into the ground nor surrounded on more than one side by walls. Creating a depression or surrounding the structure with walls can affect the flow of heated air (the draft) through the structure to a surprising degree, and this difference is

Transformative Crafts

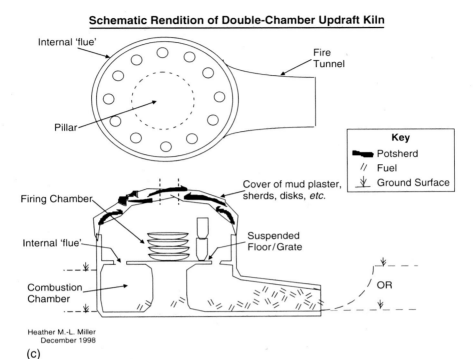

(c)

FIGURE 4.8 *(Continued)* (c) example of one type of multi-chamber firing structure, a double-chamber vertical or updraft kiln. (See Miller 1997 for full descriptions of these particular structures.)

exploited by single-chamber firing structures. *Single-chamber firing structures* comprise slightly more permanent structures, including archaeological examples referred to as "'pit kilns" or "ovens" (Figure 4.8b), where fuel is placed at the bottom or one end of the structure and the objects to be fired are on top of or next to the fuel. There might be a layer of potsherds, clay disks, or other objects placed between the fuel and the objects, but no permanent or substantial divider. This apparently minor difference can have a large effect on air flows, and is exactly the difference between single-chamber and multi-chamber firing structures. *Multi-chamber firing structures* have a separate chamber for the fuel and one or more chambers for products. Heated air passes between the fuel and product chambers, but the products are not directly in contact with the fuel, so they are more evenly fired (also see below for problems of fire-clouding). The most common sort of multi-chamber kiln is the double-chamber updraft or vertical kiln (Figure 4.8c), where the products are placed above the fuel chamber on a floor or supports. In horizontal or downdraft kilns, the fuel chamber is next to and often slightly below the chamber(s) containing the products, with a gap near the roof or vents between the chambers passing the heated air from one chamber to the next. The heated air is deflected from the roof down onto the products, hence the name "downdraft" (Hodges 1989 [1976]: 36–38). The most famous horizontal/downdraft kilns are the "climbing kilns" of East Asia, with multiple chambers built up the side of a hill (Rhodes 1968; Rice 1987: 160–161). The shape of the firing structure is also an important factor in the flow of air within the structure—whether round or square, pear-shaped or vaulted—as discussed in Rhodes (1968).

As will be apparent from these descriptions and figures, these three categories form a continuum of firing structures, of which the versions described above are only "average" examples. Repeatedly used ephemeral structures with plastered clay coverings grade into rapidly built single-chamber structures using loose mud bricks or clay lumps, both employing slopes and simple construction techniques to encourage a draft. Rhodes (1968: 20) and Rice (1987: 161) provide an illustration of the same "climbing kiln," which would be classified as single-chamber based on the design of the permanent firing structure. Yet the fuel is separated from the products by a wall made not of bricks, but of the stacked setters used to contain a portion of the objects, making this firing structure essentially a double-chamber downdraft kiln while in operation, but not after the products are removed. This sort of continuum exists even for perfectly preserved archaeological structures and modern ethnographic examples. With the added complication of poorly preserved archaeological structures (for example, usually only the base of a structure is preserved), assignment to a "type" can be a difficult task. It can even be counterproductive for investigation of production processes, if a categorization is focused too much on shape characteristics unrelated to the way the firing structure actually functioned. Classification and description of the various types of clay firing structures focused on the operating principles of these structures is generally the most helpful approach for investigations of production process, particularly interesting variations in the management of temperature, draft, and atmosphere.

The atmosphere in a firing structure has a very large effect on the final appearance of the products, as Rye (1981) and Hodges (1989 [1976]) explain. An *oxidation atmosphere* refers to an oxygen-rich firing atmosphere, one with a good draft in which all the fuel burns without exhausting the supply of oxygen available. The products are shades of red if the clays were iron-rich clays (hence the association of the pottery type "terracotta" with a red color), and white or buff or light brown with most other clays, although color chemistry of this sort is extremely complex and involves many different variables. A *reducing* atmosphere is an oxygen-poor, carbon monoxide-rich atmosphere, with a lack of sufficient oxygen to the fuel due to a restricted draft. Iron-rich clays will be fired a black or grey color due to changes in their mineral structure, as long as the reducing atmosphere is maintained. A *neutral atmosphere* is a carbon dioxide-rich atmosphere, with exactly the right proportion of fuel and air. There is often confusion over the reason why a product is a black color, as there are several different methods for achieving black-colored objects with clays of various colors. Reduction firing will fire an iron-rich clay black because the iron compounds in the clay are changed to black-colored minerals. *Sooting, smudging, or carbon deposition* can also result in black-colored products, but this is not due to an actual change in the clay minerals. Rather, with a sooty, oxygen-poor atmosphere, particles of carbon in the form of soot may be deposited in the pores of the clay objects, creating a black surface. This is often achieved by adding additional organic materials to the firing structure at the end of the firing, such as grass, sawdust, or dung. Products can also appear black, or more commonly have a black core seen in cross-section when the object is broken, because organic matter mixed in the clay is not burned out, whether because the firing was too short or too low in temperature and the clay did not properly heat all the way through, or because of a reducing atmosphere. Rye (1981: 115–118) provides a detailed description of the variations in vessel cross-sections that are indicative of various firing atmospheres and organic material content. As noted above, some waste fuels and especially dung fuels were deliberately chosen for the creation of black-fired objects, either for their high organic content (acting to scavenge oxygen) where reducing firings were desired or for their high smoke production where carbon deposition was employed. In addition, color variation can occur on the micro-level, particularly in ephemeral and single-chamber firing structures. If an object is placed so that another object covers a portion of it, that portion may not be exposed to the draft and may be reduced, while the remainder of the object is oxidized. Alternatively, if a piece of smoldering fuel is resting against an object, it may have carbon deposited in that place, but not elsewhere if the atmosphere is otherwise clean. These sorts of variation in color across a single object due to differences in firing conditions are called *fire-clouding*. Fire-clouding usually results from the masking of a portion of an object during firing, but carbon deposition is also a possibility, so care must be taken when trying to deduce firing conditions and kiln structure from the examination of fired objects, especially if only a small portion of an assemblage is examined.

    The process of firing varies with the different structure types. In multi-chamber firing structures, there is usually an opening left in the fuel chamber to add additional fuel during the initial part of the firing process (Figure 4.8c). In the other two types of firing structures, the fuel is usually placed within the firing structure and the structure sealed for the duration of the firing. Exceptions exist, such as the addition of dung over an ephemeral or into a single chamber kiln near the end of the active firing, to blacken the objects.

Ephemeral structures usually have the most rapid firing, particularly bonfire firings, which are most effective with coarse-grained or highly tempered pottery that can withstand rapid temperature changes. The inability to add fuel to most types of ephemeral and single chamber firing structures make it difficult to control temperatures in any way other than by encouraging or stopping air flow through the structure. Multi-chamber firing structures are more precisely controlled by both fuel addition and draft manipulation. Such structures tend to have a period of warming the structure rather slowly, to further dry the clay objects and remove any remaining water without breakage. The temperature is then gradually increased to a peak, where it would be held to allow penetration of heat into the products (soaking); this might happen in several stages until the maximum desired temperature is reached. Temperature decisions are usually made on the basis of the color of the objects, as the clays glow with particular colors which a potter learns by experience. The firing structure is then even more gradually cooled. Actually, gradual cooling is a concern in all types of firing structures, not just multi-chamber kilns, as too rapid cooling could result in spalling or cracking of objects.

As an example of the complexity of firing conditions, in terracotta pottery where an even firing of the desired colors needed highly oxidizing conditions, potters could maintain a strongly oxidizing atmosphere via a good flow of air through the firing structures (i.e., a strong draft). This could be achieved with several different firing structure designs, but if the potters also wished to avoid fire-clouding, both needs could be met by the physical separation of products and fuels in a double-chamber updraft kiln or a downdraft kiln. While providing enough fuel and enough of a draft for a uniform oxidizing atmosphere, the potters also had to carefully control the firing temperatures, especially if firing near the sintering point of a clay. If the sintering point was near the vitrification point, a very narrow temperature range had to be consistently reached; for example, firing the products at temperatures of 800–850°C and yet not shooting up above 1000°C and melting both the products and perhaps the firing structure itself. Temperatures as well as atmosphere would have been affected by the amount, type, and rate of addition of fuel, which is an additional reason why the analysis of these fuel materials is so important. Temperatures are also controlled by managing the flow of heated air (the draft) through the firing structure, via the design of the structure as well as the use of even such simple kiln furniture as sherds and clay disks. Draft is thus an important consideration for both atmosphere and temperature control during firing.

These firing structure types do not represent a chronological sequence where increased separation of the fuel from the products solves all problems and is the preferred method once invented. On the contrary, I have illustrated quite a different process occurring in prehistoric South Asia, where new, increasingly complex firing structures were developed over time, but variants of the older types continued to be used at the same time (H. M.-L. Miller 1997). Why this is occurring is one of the more interesting questions about the choices made in pottery firing. Did craftspeople choose the type of firing structure to use based on the quality of the objects being produced? Or the quantity? Or the size of the objects? Or the fuel costs? To contrast the costs of a double-chamber updraft kiln with an ephemeral, "open-air" firing structure, the updraft kiln requires more labor to build and maintain, requires the permanent dedication of space to firing (an important issue in urban contexts), and usually requires more fuel. The advantages of an updraft kiln are that the firing can be more easily controlled and is less affected by weather conditions, higher temperatures can be reached, and the products are protected from the smoke and dirt of the fuel. An archaeologist has to look for information on which of these issues were important for the potters they are investigating.

For example, potters in South Asia today primarily use the more "primitive" of the types of kilns used in the past, either ephemeral or simple single-chamber firing structures. (See Sinopoli (1991) and Rye (1981) for examples.) They have made this choice not because they are not aware of the existence of multi-chamber firing structures, but because they are more concerned with fuel consumption than achieving a uniformly oxidizing kiln atmosphere and sintering temperatures. On the contrary, they do not want to fire their products to a sintered state. They are not making the elaborately painted terracotta assemblages of serving vessels like those that the ancient Indus civilization potters fired in updraft kilns. Glazed wares and porcelains replaced these terracotta serving vessels in the historic and modern periods, many imported from Western Asia, Europe, or East Asia. Glass, metal and plastic vessels have also replaced many types of pottery vessels. Instead, the main product of many modern South Asian potters is coarse terracotta water jars. Consumers prefer these water jars because the permeability of the terracotta vessels allows water to transpire from the exterior surface and keeps the liquid on the interior cool. Therefore, firing at high enough temperatures to decrease this permeability would be a disadvantage. These vessels are sometimes painted (Figure 4.7), and modern potters in South Asia claim that application of a sandy coating also helps to keep water cool (Dales and Kenoyer 1986: 42; Rye and Evans 1976: 53). However, a slightly fire-clouded surface is not a major concern, and the cost of fuel definitely is an issue. The ability to use waste fuels and dung fuels is a major economic advantage for a potter. For all these reasons, ephemeral or simple single-chamber firing structures are preferred. Similarly, the firing skills of the prehistoric potters of Iranian and Pakistani Baluchistan should not be underestimated because they used simple single-chamber firing structures. Wright (1989a; 1989b) has demonstrated that the painted greyware vessels fired in these structures were subjected to very complex cycles of reduction and oxidation to achieve their final colors. Such cycling of atmospheric conditions was likely much easier in these single-chamber kilns than in double-chamber kilns, especially the reduction cycles. If so, these simpler kilns would be the more appropriate tools for this more complex firing system.

POST-FIRING SURFACE TREATMENTS AND SECOND FIRINGS

Most unglazed fired clay objects are finished at this stage, although some have additional post-firing surface treatments carried out by the producer, or even by the consumer. Post-firing surface treatments can be functional or decorative (Rye 1981; Rice 1987; May and Tuckson 1982; B. E. Frank 1998). In many places, pottery is 'seasoned' by the potter or the consumer to prepare it for use. It might be lined with resins or oils to decrease the permeability of liquid storage containers, or filled with liquid foodstuffs and heated prior to a cooking vessel's use to seal the inner surface and prevent sticking without compromising taste. Seasoning may also be done to increase strength. Vegetable dyes or other colorants that cannot survive firing are also applied as post-firing decorations.

An additional firing stage might also be required for glazed objects. Glazed wares were sometimes fired once prior to application of glazes, if the bodies and glazes were not matched in temperature and atmospheric requirements. If two firings were needed, the object was shaped and dried, then fired in the first *biscuit (or bisque) firing*. Painting and glazing was done, then the final firing took place. Glazed wares and porcelains often had far more complex stages of production than are outlined here, as described in Hodges (1989 [1976]: 42–53, see especially the production diagram on p. 52), Rye (1981), Rice (1987), Henderson (2000), and other specialist works, such as the examples in the edited volumes of the *Ceramics and Civilization* series produced by the American Ceramics Society. Glazed ware production can be similar to some of the types of ceramics discussed in the next section, the vitreous silicates, so I will discuss their production in the next section as well.

VITREOUS SILICATES: GLAZES, FAIENCES, AND GLASS

Glazes were discussed in the previous section, as a surface treatment for fired clay objects. They will also be included in this section, as a vitreous silicate applied over a body made of clay, stone, or in the case of enamel, metal (Hodges 1989 [1976]). The vitreous silicates discussed in this section, glazes, faience, and glass, are made from essentially the same raw materials and can have very similar compositions. These categories overlap, which has particularly been a problem for categorization of the faiences. By *faiences*, I refer to the range of (primarily) soda-lime-silica vitreous materials known from ancient Egypt, Mesopotamia, the Indus Valley, and later Europe, which were modeled like clay but were quartz-based rather than clay-based. These materials were primarily used to make beads and other ornaments, as well as figurines, small vessels, inlay pieces, and other relatively small objects. My definition of "faience" focuses on the working properties of the material, as well as the composition. This contrasts with Moorey's (1994:167) definition of faience as "a composite material consisting of a sintered quartz body and a glaze," and the more narrow definition of faience used by most Egyptian researchers as a soda-lime-silica vitreous material having distinct glaze and body layers (summarized in Nicholson 1998; and further refined in Nicholson 2000; also see Shortland 2000). I do not object to these definitions; as will become apparent, however, they can be very problematical for categorizing the wide range of objects found in all regions of Eurasia. Given the very small percentage of objects subjected to compositional analysis, definitions requiring this sort of information are difficult to employ, as Moorey (1994:168) also notes. A great many more faience objects have been found and a good deal more analytical research has been done in Egypt than in most other regions on these types of vitreous materials, allowing very fine-scale divisions that are not always as clear elsewhere. While this careful characterization on the part of Egyptian researchers is essential to untangle the development and diversity of these vitreous materials, the less complex terminologies used elsewhere and an additional focus on working properties are more suited for my purposes here. Finally, the main distinction between faience and glass is that faience is only *sintered*, heated so that a portion of the constituents melt to form a fusing agent to hold the remaining unmelted materials together. *Glass* ingredients are completely melted into a liquid that fuses on cooling (Moorey 1994:167).

In spite of the similarity of raw materials, the goals of the craftsperson and the problems that needed to be solved were quite different between these crafts, so that there often are some significant differences in the production choices made. I will highlight both parallels and differences throughout this section. As noted in the introduction to the chapter, all these vitreous silicates were only developed in Eurasia, not in Africa or the Americas (Rice 1987: 20), which adds further weight to the suspicion that they are intertwined in their technological development. Both Hodges (1989 [1976]) and Henderson (2000) have very informative chapters on glazes and on glass, although both only mention faience production in passing. Several of the major texts referenced in the Fired Clay section also contain information on glazed clays (Rice 1987; Rye 1981). Frank's (1982) summary of early glass manufacture is still a useful starting point. Nicholson (1993; 1998; 2000; Nicholson and Henderson 2000) has written several key summaries of Egyptian faience and glass production. Moorey (1994) covers all of the vitreous materials for Mesopotamia, as well as most of the other crafts discussed in this volume. Michael S. Tite and Pamela Vandiver have done a great deal of archaeometric work on faiences and other vitreous materials over the past few decades, with numerous publications in *Archaeometry*, the series *Materials Issues in Art and Archaeology*, and other notable venues. The *Journal of Glass Studies* is a major source for articles on ancient and historic glass production, as is Kingery and McCray (1998), and the extensive bibliography in Henderson (2000). Fleming (1999) supplies an absorbing overview of the use of glass in Roman life, while McCray (1998) examines Renaissance Italian glass production from an archaeological, historical, and technical perspective, interweaving data from all these datasets and providing a useful entry into technological approaches in the history of technology. Glazes and glasses are still produced and have been ethnographically studied, and there are historical texts detailing their production in the past. In contrast, faiences have not been produced in modern times and there are few historic accounts of their production, further complicating our reconstructions of their production processes.

The general production of vitreous silicates (glazes, faiences, glass) employs the following stages (Figure 4.9):

1. Collection of silica, fluxing materials, any needed colorants, and fuels
2. Preliminary processing of silica and of fluxes and colorants (crushing, sieving; burning of plant ash)
3. Creation of the faience body, glaze, and glass mixtures, in some cases including fritting; for glasses, melting (glass making)
4. Shaping of faience and glass (glass working) using modeling, molding, blowing, and numerous other methods for glass
5. Application of glazes to faience or glazed objects
6. Firing of faience and glazed objects; annealing of glass
7. Possible surface treatments.

Collection and Preliminary Processing

Glass, glazes, and other vitreous materials are technically classed as liquids at normal temperatures, albeit with an extremely high viscosity, because glass molecules have a random, non-crystalline structure, unlike most solids. This does not mean that these vitreous materials behave as liquids at room temperature; all of these materials function very much like solids unless heated. McCray (1998: 35, ftnt 8) presents a clear explanation of the structure of glass for the non-specialist, including the debunking of the common idea that old windows have 'flowed' because glass is liquid-like. Henderson (2000: 24–25)

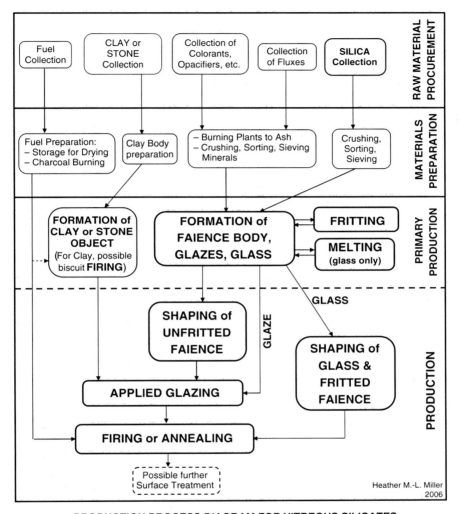

PRODUCTION PROCESS DIAGRAM FOR VITREOUS SILICATES

FIGURE 4.9   Generalized production process diagram for vitreous silicates (greatly simplified).

provides a more extensive technical definition. The "random network" or "open network" pattern of their molecules at room temperature allows the vitreous silicates to accommodate many other atoms in their structure, such as the variety of metallic atoms that give the vitreous silicates their wide range of colors. The base raw material for all the vitreous silicates is, not surprisingly, silica. Silica is very widespread geologically, and most sand deposits are primarily composed of silica. However, most sand deposits contain considerable impurities that would affect the colors and properties of the desired product, so that significant effort seems to have been made to find relatively pure deposits of silica sand, the most famous such deposit during the Roman period being that of the Belus River in Syria. Alternatively, quartz pebbles and rock crystal were crushed to get quite pure silica, but with much more labor. Flint was also crushed to create glass in Europe, since the sands available contained too many impurities to make a clear glass (Hodges 1989 [1976]: 54).

Because silica melts at a very high temperature for pre-industrial kilns, 1710°C, *fluxes* or *modifiers* need to be added to lower the melting point of the silica. These fluxes are a variety of metallic oxides, which include oxides of sodium, potassium, lead, magnesium, and alumina (Hodges 1989 [1976]: 44–45). Many of these oxides are found together in any particular glaze or glass mixture, as many of them can come from the same source (e.g., plant ashes or mineral sources), or are found as impurities in the silica source or in the clay or stone body of glazed materials and mix with the applied flux during production. Thus, glazes are usually not characterized by the presence or absence of a particular oxide, but by the relative proportion of the oxides present. Each type of flux compound or modifier had different properties that would be an advantage or disadvantage for the production process or the final object, such as high plasticity or low melting point during production, or increased luminosity of final object. For use, the mineral fluxing agents would need to be crushed or ground, sorted and sifted; some were also burned prior to crushing. To create plant ash fluxes, particular plants were burnt, and the ashes collected and cleaned.

Glaze and glass classifications are frequently based on the type of flux (modifier) or its source of origin (e.g., Hodges 1989 [1976]: 48–50, 56; Rye 1981: 44–46; McCray 1998: 36). There are two main divisions of vitreous silicate materials based on their fluxing systems: lead and alkaline. Both of these types apply to glass and glazes, while faiences mostly employed alkaline fluxes. Hodges adds a third type of glaze, slip glazes, which employ an iron-rich clay slip that is vitrified (melted to a vitreous state). Clay slips are mentioned here primarily to note that they are frequently confused with glossy, unvitrified slips, which are not glazes, leading to incorrect statements about the presence of early glazes (e.g., Rice 1987: 20). Lead oxide fluxes are made from various metallic and mineral sources of lead, which were burned and powdered when used for glass production (Henderson 2000). As a glaze, lead oxide compounds could be applied either directly as powder, or more commonly suspended in water (Hodges 1989 [1976]). Alkaline vitreous materials, employing alkaline fluxes, are divided into soda-lime types, potash-lime types, salt glazes, and feldspar-lime or felspathic glazes. The last two types are found only in glazes, and are produced only at high temperatures (above 1100°C). Salt glazes are formed in an unusual way, and are found in Europe and North America after the twelfth century ad/CE (Rye 1981; Hodges 1989 [1976]). Rather than mixing the sodium chloride fluxing agent (rock salt) with silica and applying it to the surface of the object, as is usual, salt is thrown into the fire during firing to produce a sodium oxide vapor that combines with the silica in the clay body of the objects and forms a surface glaze. In feldspar-lime glazes, feldspar minerals and often a source of lime flux (calcium oxide) were ground, mixed with silica, and applied as usual (Rye 1981; Hodges 1989 [1976]). Feldspars contain alumina, which functions both as a flux and as an *intermediate*, strengthening the glaze to prevent crazing and making it more viscous ("stiffer") so the glaze does not run. These very hard, high-temperature glazes were primarily applied to porcelain bodies, and Rye notes that feldspar glazes were limited to East Asia until the eighteenth century ad/CE.

The most common alkaline vitreous silicates are the soda-lime types and potash-lime types. There are both mineral and plant sources of the sodium oxide and potassium oxide fluxes, including sodium carbonates (soda), potassium nitrate (saltpeter), and a range of plant and wood ashes. In addition, a source of lime (calcium oxide) was necessary for a stable vitreous material; without calcium oxide, alkaline vitreous materials would dissolve in water. Calcium oxide, itself a flux, comes from limestone, chalk, gypsum, and many other minerals, as well as bone ash and even plant ash, and may be present in either the fluxing agent or the silica source in sufficient quantities that deliberate addition is not necessary, depending on the product. Sodium-rich glasses and other vitreous materials can be further differentiated by whether mineral or plant sources were used for the fluxing agent (McCray 1998: 36; Henderson 2000: 25–26). An important mineral source of alkaline flux used in ancient, Hellenic, and Roman faience and glass manufacture was the mineral natron, found in Egypt and famous for its use in mummification. Natron is a mixture of sodium compounds, so that glass made with this flux has high sodium and low potassium and magnesium content. However, the most widespread source of alkaline fluxes used for glazes, faiences, and glass were plant ashes. The plants used from the Mediterranean to western India were typically desert bushes, often *Salsola, Suaeda,* or *Salicornia* species, containing not only sodium oxides from soda ash but also significant quantities of potassium oxides, magnesium and phosphorus. (*Salsola* will be familiar as the introduced Eurasian "tumbleweed" that colonized the western deserts of North America.) Rye (1976) describes the processing of several common soda ash-producing desert plants of the wider *Chenopodiaceae* family, to obtain ashes for the production of glazes and soap. In contrast, McCray (1998) notes that glass making in parts of Europe primarily employed wood ashes with a high potassium content and very little sodium. Finally, while all the ancient faiences that have been tested made use of alkaline fluxes, the flux magnesium oxide, in the form of talc/steatite fragments, was also a major component in one of the Indus faiences, "steatite-faience," although its presence may relate to ideological rather than functional reasons (see below and Chapter 6).

The last set of materials added to all the vitreous silicates were used in tiny amounts but had a major effect on the final product: colorants and opacifiers. The chemistry of the colorants and opacifiers used in glasses, glazes, and faiences are very complex. In practice often the glass or glaze maker's biggest problem was to exclude traces of color to create a clear glass, which has been aided historically through the addition of small amounts of de-colorants. *Colorants* are metal compounds, and the colors they produced can depend on the proportion of the colorant present, the firing temperatures reached, what fluxes are used, and whether fired in oxidizing or reducing conditions (Hodges 1989 [1976]: 45–46; Rye 1981: 47, Table 4.2; McCray 1998: 37; Henderson 2000: 29–38). Moorey (1994: 184–186) also summarizes the analyses of colorants for vitreous materials in the Near East, and explains some of this complexity. For example, the common turquoise blue color found in faiences, glazes, and glasses across Eurasia in many time periods is produced by copper colorants fired under oxidizing conditions with an alkaline glaze. Copper colorants fired under oxidizing conditions with a lead glaze produce an emerald green color, while copper oxides fired under reducing conditions produce a ruby red color. However, if there is more than 5% copper colorant present, the color darkens to black. Typical colorants used include copper, cobalt, iron, and manganese. *Opacifiers* are metal compounds that create opacity in the vitreous materials; tin and antimony compounds were the most common ancient opacifiers, and could be mixed with various colorants to produce opaque rather than translucent colors for glass and glazes.

There is some debate over whether additional *binders* were added to faiences, as was done with glass pastes (Hodges 1989 [1976]). These binders would be adhesive organic materials (gum, honey, oil, etc.) used to help shape the objects while still wet, and which would burn out during firing leaving no trace. Small amounts of clay binders have also been suggested, and while clays should leave trace elements, such traces might be very difficult to distinguish from typical impurities found in silica and flux sources. Both of these problems have vexed archaeologists trying to analytically and experimentally reconstruct these materials, and I know of no case where definitive analytical evidence has been found for either clay or an organic binder in faiences, although both have been used in experimental attempts at re-creation (e.g., Nicholson 1998, 2000).

A special addition to one type of faience from the Indus Valley has been identified: talc (steatite) fragments. Talc is thought to have had a special place in Indus cosmology, so that talc beads, seals, and other objects had particular social and perhaps ritual meanings, as discussed in Chapter 6. Talcose faience, or "steatite faience" appears to be a uniquely Indus material, not found elsewhere (Barthélémy de Saizieu and Bouquillon 1997: 75). Talcose faience, at least that found at the sites of Mehrgarh and Naushara in the Baluchistan hills, was composed of talc fragments "embedded in a fine matrix made of talc, flux elements and a colouring agent (copper oxide)" (Bouquillon and Barthélémy de Saizieu 1995: 50). Aside from this difference in composition, it appears to have been produced in the same ways as purely siliceous faiences. The work by Barthélémy de Saizieu and Bouquillon has provided a highly informative chronological sequence of bead types from Mehrgarh/Naushara that seem to reflect stages in the development from the manufacture of glazed massive talc beads to siliceous faiences. (See Figure 6.3 for a schematic of fired talc, glazed talc, and faience development for the Indus region.) Talcose faience has provided a clear link between the talcose and vitreous siliceous industries in the Indus, but has also made the discussion of Indus talcose and vitreous siliceous materials even more complex. It has also made it difficult to use terminologies developed for other regions (e.g., compare the very different use of terms in Moorey 1994:167-168 and Miller in press a). It is very difficult, perhaps impossible, to visually distinguish talcose faience beads from siliceous faiences and even talc beads, so the range and number of objects created from this material is difficult to judge, but talcose faience has been analytically documented at the urban site of Mohenjo-daro as well as the village sites of Mehrgarh and Naushara. The modern analyses of materials from Mehrgarh and Naushara in Baluchistan were all done on beads, but the two analyzed objects from the older excavations at Mohenjo-daro were a human figurine fragment and the base of a small vessel (Mackay 1931: 576), indicating that talcose faience was used for the same variety of objects as the siliceous faiences. Furthermore, Mackay (1931: 576) and Barthélémy de Saizieu and Bouquillon (1997: 67-68) clearly state that these objects, found on analysis to be talcose faience, were indistinguishable from siliceous faiences on the basis of visual examination. Much of the "faience" identified at many Indus sites may well be talcose faience rather than siliceous faience, and it will be interesting to see the chronological and spatial patterns of talcose faience production with further systematic analytical research at additional sites. The social information which may be encoded in such objects is further discussed in Chapter 6.

CREATING THE VITREOUS SILICATE MIXTURES; FRITTING; MELTING OF GLASS (GLASS MAKING)

As Hodges (1989 [1976]: 54) emphasizes, the various vitreous silicates used similar raw materials, but the working qualities and goals of the various types were quite different, so that different proportions of materials were used. Very different issues faced craftspeople glazing stone materials or faience bodies and those glazing clay bodies (pottery). Craftspeople even in the same industry and same region seem to have used multiple methods to attain products very similar in appearance, as the variety of faience production methods illustrates. Some of these variations are chronological, some may represent regional techniques (perhaps even "schools" of production methods or lines of apprenticeship), while some variations may simply represent expedient use of the materials available at any given time.

Beads made from magnesium silicates (primarily talc/steatite) and silicate (often quartz) appear to be the earliest glazed materials, with glazed faience found soon after, dating from at least the fifth millennium BCE in Mesopotamia, Egypt, and the Indus Valley (Moorey 1994; Nicholson 2000; Barthélemy de Saizieu and Bouquillon 1997). It is noteworthy that the glazing of clay objects develops long after the glazing of stone and the production of faience in all these regions (Moorey 1994). Stone objects to be glazed were previously formed by knapping, abrading, cutting, or the other reductive processes described in Chapter 3 for stone working, while clay objects were formed, dried, and in some cases fired prior to glazing, as noted in the Fired Clay section (Figure 4.9). Like these other glazed objects, most of the faience materials described below were also composed of a body to which glaze of a different composition was applied. A major challenge for the artisan was insuring the binding of the glaze to the body, both before and after firing. Wet glazes had to sufficiently adhere to the body and not drip off, but yet not be absorbed into the body completely on drying. The fired glaze had to be sufficiently bound to the body so it did not flake off, but excessive shrinkage of the glaze had to be avoided or the glaze would crack or craze. Each type of body had different characteristics, so the glaze composition and/or the firing regime would have had to be a little different for each. Faiences could also be self-glazed, as discussed below, removing this problem of glaze-body fit for the producer. However, faiences have the same problem as glasses, which is that they are not supported by a body or backing, but must be self-supporting to form objects. Glass makers had great concerns about transparency of the material, which was less of an issue for glazes and of little concern for faiences. Glass and faience workers desired mixtures that would remain plastic during working, while glaze workers might want the glaze to set relatively rapidly.

To create all of these vitreous silicates, the finely crushed silica, modifier (flux or fluxes), colorants, and any other materials would be mixed together. Almost all faience body mixtures were simply combined prior to forming of objects, with a little water added to aid in forming. Some glaze mixtures could also be directly applied to a stone object, a faience object, a dried clay body, or a biscuit-fired clay body, usually suspended in water to aid even application. The clay bodies might be previously painted with colored pigments or not; Hodges (1989 [1976]: 47–48, 52) discusses the complexity of painting, glazing, and firing regimes for glazed clay, and provides a schematic production diagram for various alternatives. In addition, both soda-lime and potash-lime glazes were usually fritted before application to a clay body, because the high solubility of these substances would result in their leaching into the clay body, even for a biscuit-fired object. Hodges suggests that this was why soda-lime alkaline glazes were used only to glaze quartz and talc stone and faience (itself a soda-lime-silica mixture) in ancient Egypt, and not pottery. Depending on how one defines the term faience, a few types of faience were also fritted prior to forming of objects, as were all glasses.

Fritting is the process of heating the finely ground mixture of silica and fluxes, and sometimes the colorant, to the sintering or fusing point but below the melting point, usually while raking or stirring the mixture. A fritting temperature of around 650–800°C is used for many glasses. The resulting frit is then reground into a very fine powder, with a high surface to volume ratio. Fritting ensures proper mixing of the materials, removes impurities and gases, and creates this fine fused powder that facilitates the next stages: object formation for fritted faience, application to object for fritted glazes, and melting to create glasses (Figure 4.9). A few faiences were made using a fritting stage, and fritted vitreous silicates that were hand-formed or molded like clay are reported from the Indus (McCarthy and Vandiver 1991), Mesopotamia (Moorey 1994), and Egypt (Nicholson 1998: 55; Nicholson 2000: 177–178; Nicholson and Henderson 2000: 205), although sometimes under different names such as "Egyptian blue frit." (Researchers in Egypt use a "splitting" approach to vitreous silicate materials, while those in Mesopotamia and the Indus use a more "lumping" approach, probably in part due to the relative amounts of analytical work that has been done in these three areas and the relative amount of material in collections and available for analysis.) For example, McCarthy and Vandiver (1991) analyzed an extremely strong, smooth, non-porous type of siliceous faience from the Indus which had been fritted. Multiple stages of fritting and regrinding may have taken place to produce a particularly fine material. The objects produced from this material would be very homogeneous, and so stronger, allowing the production of such structurally precarious objects as bangles (McCarthy and Vandiver 1991). This fritted form of siliceous faience and the talcose faience described above indicate the tremendous experimentation in vitreous material production taking place in the Indus, as discussed in Chapter 6. Similar degrees of experimentation with this continuum of vitreous silicate materials also occurred in the other regions of the ancient world (e.g., Moorey 1994: 169).

For *glass making*, frit is then *melted* at much higher temperatures than the fritting stage, at least 1100–1350°C depending on the mixture, and a piece of old scrap glass (*cullet*) of a similar composition is usually added to the frit to physically assist the melting process. Any colorants, opacifiers, or other agents are typically added once the melt is underway. The uncolored glass might even be melted once, then the raw glass crushed and sorted to select for high-quality raw glass for a second melting with the addition of color (Rehren, et al. 2001). Melting is the final stage in the production of glass material, but only the beginning for forming glass objects. Like metal ingots, raw glass or *glass ingots* once produced could be traded and re-melted, "alloyed" with new colorants if clear, formed, and recycled as scrap glass or cullet, although to a limited degree in comparison to the almost infinite recyclability of metal. Thus, finding melted glass debris or tools at an archaeological site (e.g., Figure 4.10a) can be a sign of glass melting, re-melting, *or* working, and alone is not necessarily an indicator of glass making from initial raw materials. Rehren et al. (2001) discuss such an example of the separation of glass making and glass working production sites in New Kingdom (Late Bronze Age) Egypt. Glaze mixtures could also be traded and applied to pottery at another location, but this was not as common as the widespread ancient trade in glass.

SHAPING OF FAIENCE AND GLASS OBJECTS

The faience bodies described above would be wetted and formed, shaped primarily either by hand or in a mold. The working properties of faiences are very different than clays, becoming soft and flowing as it is wetted and shaped, but cracking if shaped too rapidly (Nicholson 1998, 1993). Experimental re-creations have indicated that faience bodies appear to have better working properties if made with more finely ground materials or with the addition of some binders. Modeling of faience objects would nevertheless be a very different experience from modeling clay. With care, surface details could be carved into the object after drying, although Nicholson (1998: 51; 2000: 191) points out that for faiences glazed by efflorescence (below), carving the surface after the object has dried will remove the glaze from that area. Freehand modeling was used to make many small figurines and ornaments such as beads and bangles; beads may also have been modeled around sticks or rods to create a perforation. Small faience vessels and other objects were sometimes formed around sand-filled cloth bags or straw forms, based on marks left on the interior of some vessels. Molds were extensively used to shape faience objects in Egypt and to a lesser extent Mesopotamia, but seem to have been rarely used in the Indus where far fewer figurines and inlay pieces were produced. While faiences would not impress as well as clays, the use of molds would allow greatly increased production of objects. Faience pieces could also be easily combined into composite objects after drying, more easily than for clay (Nicholson 1998). Although molds can be made from a variety of materials,

Transformative Crafts 139

(a)

(b)

FIGURE 4.10 (a) Glass melting debris and (b) glass molding debris.

clay, wood or metal among others, primarily single-sided fired clay molds have been found in Egypt. There is evidence for multi-part molds, however, and Nicholson also points out that finished faience objects can themselves be used as molds. Finally, Nicholson (1998: 52; 2000: 189) notes that in late periods, definitely by the Greco-Roman period in Egypt, faiences with a high proportion of clay in their bodies were even thrown on the wheel. This material again shows the difficulty of defining faience solely on the basis of composition. It is a very interesting material in terms of showing the overlaps between all these vitreous silicate crafts—a major point of this volume. Similar sorts of "fritware" or "stonepaste" are described by Henderson (2000: 181) for historic Islamic pottery from Western Asia, as a material made from "crushed silica combined with a small amount of clay and crushed glass." These separate inventions of similar materials provide an excellent example of the complex, intertwinning and repeating trajectories of invention, innovation, and trade seen in the history of the vitreous materials.

Glass working also has made use of hand-shaping and mold use. Historically, glass pastes, glass ground to fine powder and mixed with an organic adhesive, have been shaped by hand, in molds, or even thrown on the wheel (Hodges 1989 [1976]: 57). The line between such glass pastes and fritted faiences is very thin from the perspective of forming, although the difference in materials (no fluxes were mixed with the glass pastes) would result in very different firing requirements. A variation on mold use would be the *casting* of molten glass, similar to the casting of molten metal (Moorey 1994: 206). Other than cast glass ingot production, however, forming of glass in molds prior to the modern period has primarily taken place in association with blowing, as described below. The remaining major methods of glass shaping can be categorized as abrading, cane (rod) formation, core-dipping and core-winding, and blowing (Hodges 1989 [1976]; Moorey 1994). Marvering is often employed as a secondary shaping step for the last two shaping categories. *Abrading* or cold-cutting must have been rarely used as a forming method except perhaps for beads, as this process simply treats glass as stone and shapes it by the abrasion techniques discussed in the stone section of Chapter 3, losing all the special advantages of glass as a pliable working material. However, abrasion was a fairly common method of post-firing surface treatment for all the vitrified silicates. Glass *cane* can be produced in short lengths by slowly pouring a measure of glass that cools as it falls. For longer lengths, a method similar to taffy-pulling is employed, with one end of a mass of glass gathered from the melt and stuck to a metal plate on a wall, and the other end pulled out on the end of a metal rod by walking away from the wall. In *core-dipping*, a core of clay or fabric-wrapped sand attached to a rod is immersed in molten glass, or heated and rolled in powdered glass, to coat the core (Nicholson and Henderson 2000: 203). *Core-winding* involves the winding of heated drawn rods of glass (canes) around a core. In both of these core-built processes, the entire object is then heated and rolled on the flat smooth surface of a marver to produce a smooth outer surface. *Marvering*, rolling the hot glass on this object, was also done with blown glass, to shape, cool and smooth the surface. *Glass blowing* is the premier method of glass shaping, and can be sub-divided into two types, free-blowing and blowing into molds. After gathering molten glass onto the end of a metal tube, the glass worker could *free-blow* the glass into a hollow shape, with perhaps additional shaping with another iron rod or the marver, or *hot-working* to add handles or an applied glass decoration. Free-blowing requires great skill, and it can be very time-consuming to make certain shapes, although it allows a wide variety of shapes and designs. For the more rapid production of standardized shapes, glass is *blown into molds* of one or more pieces, which might also be used to form the glass into decorative shapes or surface designs (Figure 4.10b). Bottles have been made by blowing into molds for centuries, with the characteristic base a product of manipulating the end of the bottle on a second rod (Hodges 1989 [1976]: 58, Fig. 6). Finally, some of the most complex decorative techniques used in glass working might take place after the constituent glass canes, threads, and vessels were completed. These would all require a second stage of heating. To produce *mosaic glass*, various pieces are arranged on a plate or mold and then heated until fused (Moorey 1994:204-205). Glass designs can be *inlayed* into glass objects, often using glass threads or canes, by applying the design then reheating and marvering the entire object. This type of inlaying with reheating was used in a variety of techniques, including the process of millefiore production. Hodges succinctly explains these and other decorative glass working techniques, as well as the production of enamel, another related craft where vitreous material is fused to metal. Summaries of the literature on these techniques for Mesopotamia and Egypt can be found in Moorey (1994) and Nicholson and Henderson (2000).

APPLICATION OF GLAZES TO FAIENCE AND OTHER MATERIALS

As discussed in the Fired Clay section above, unfired or biscuit-fired clay objects could have glaze mixtures applied by dipping the object into the liquid mixture, less often by brushing the liquid mixture onto the surface, or even applying the glaze mixture as a powder as Hodges (1989 [1976]: 49) describes for lead glazes. The same techniques were used for glazed objects made of talc or quartz stone. Glazes could similarly be applied to faience objects by dipping the faience into a liquid mixture or applying the glaze as a powder, but the nature of the faience body allowed for other variations as well. Moorey (1994) and Nicholson (1998, 2000) classify the varieties of faiences primarily on the basis of their glaze application techniques, as deduced from laboratory and experimental studies. Following Vandiver's pioneering work, these methods of glazing are (a) a body with a separately applied wet glaze; (b) a body glazed by cementation in a glaze powder; and (c) a body self-glazed by efflorescence of materials from within the body. Faience varieties are described for Mesopotamia and Egypt by Moorey (1994: 182–186) and Nicholson (1998; 2000), who summarize the extensive work by Vandiver, Kaczmarczyk and Hedges, Tite, and others. The much smaller body of work for the Indus is summarized in Miller (in press-a).

For an *applied wet glaze*, the faience body would either be painted with or dipped into a separately manufactured, colored, liquid glaze, then fired. As with glazed clay, some faiences with applied wet glazes might have been fired prior to glazing (Nicholson 1998: 51, ftnt 16), then fired again with the glaze. In the process of *cementation*, glaze is "applied" by embedding the body in a dry powdered glaze mixture, and firing it in this powder. Cementation is often incorrectly referred to as "self-glazing" (like efflorescence, below) because no wet glaze mixture is applied prior to firing. However, the glaze is still a separate material applied to the body; it just adheres to the body during the firing. Cementation works best on bodies containing at least some silicate, allowing a bond between the silicates in the body and the glaze powder as they both are heated. The silica in the object body and the glaze are both "wet" at high temperatures (around 1000°C), but the lime or other unreactive material in the glaze powder are not, so the objects do not stick to the bed of powder and no setters or nonstick surfaces are needed. The presence of lime or some similar material in the glaze powder is thus crucial. Cementation also removes many of the difficulties of glaze-to-body adherence inherent in wet glaze application. Finally, some of the faiences were truly "self-glazing"; that is, a separate glaze was not applied, but formed from the migration of materials within the body of the faience (Moorey 1994). In this method of *efflorescence* glazing, alkalis (usually from plant ash) within the body of the faience migrate to the surface during drying of the body, and precipitate or effloresce out to form a powdery layer. The drying stage is thus very important, and the faster the drying, the thicker the glaze coat. During firing, this layer fluxes the silicates in the surface of the body, and creates a glazed surface. Any desired colorants are thus included in the body of the object, not added in the glaze, so the glaze and body are the same color. This method also avoids the glaze-to-body adherence problems of a wet glaze application method, although care must be taken when handling the dry object prior to firing or glaze will be removed from the surface, as discussed under Shaping above. The problem of objects sticking to their setters during firing must also still be solved for efflorescent methods of glazing.

Firing of Faience and Glazed Objects; Annealing of Glass

All of the vitreous silicate objects were fired to temperatures of 800–1000°C or higher at some point in their production process, sometimes at several points. There were thus a range of complex firing structures and firing tools used in these vitreous silicate crafts about which we still have far too little information, especially for the faiences. The types of firing structures used to fire glazed clay objects are similar in most cases to firing structures used for unglazed clay objects that needed similar firing conditions, as discussed in the Fired Clay section above. However, glazed clay objects also employed a range of specialized kiln furniture, principally containers or *saggars* to keep the objects protected from exposure to flames, fuel, and soot, and *setters* to separate the glazed objects and keep them from sticking to each other or the firing structure and containers when stacked in the kiln. Rice (1987), Hodges (1989 [1976]), Rye (1981), Rhodes (1968), and particularly Henderson (2000) describe and illustrate these structures and tools for glazed pottery and clay object firing. There is very little published on firing for other sorts of glazed objects, such as glazed stone; it seems to be generally assumed that firing methods would be similar to those used for early faience production. The faiences were likely fired in various types of structures during their long existence across such a large region, but their firing structures and other firing tools have been surprisingly elusive. There is considerable discussion in the literature for all regions about the likelihood of temporary firing systems, such as containers that were fired in an open structure or even a bonfire. The most thoroughly investigated and published faience firing assemblages have been discussed by Nicholson (1998; 2000), who has excavated some of these materials and structures himself in Egypt, together with associated glass working assemblages. Miller (in press-a) summarizes the small amount of data for faience firing for the Indus. Finally, Moorey (1994: 202-203) details the evidence for glass firing in Mesopotamia, while Henderson (2000) provides an extensive discussion of glass melting furnaces and other firing structures, ranging widely across time and space. Glass making and working requires a variety of stages where high heat is applied to the materials or objects, so glass kilns that were designed to be used for most of these stages might be very complex, as is illustrated so well in Agricola (1950 [1556]). Glass production typically has two stages of firing to the molten state, the initial melting for glass making and a re-melting for glass working, and two stages of firing at lower but still high temperatures, the fritting stage prior to glass making and the final annealing stage of the shaped object. Newly shaped glass objects had to be *annealed*, that is, held in a heated condition and only slowly cooled to prevent cracking or breaking from sudden temperature changes, so glass working furnaces typically had annealing ovens where finished objects could be left until ready for any post-firing surface treatments.

Post-Firing Surface Treatments

Glazed, faience, and glass objects might be lightly polished, ground or smoothed after the final firing or annealing, to remove any traces of forming methods such as molding or any slight defects on the surface. These abrading techniques, as described in the Stone section of Chapter 3, can involve rubbing with fine-grained hard materials, such as sandstone or siliceous leaves. Abrading also includes rubbing with a soft material like cloth or leather plus an abrasive powder, and sometimes a liquid or oil to spread the abrasive and decrease heat. The latter method of abrading or polishing, rubbing with a soft material and an abrasive powder, was the most likely to have been used for vitreous silicate objects, given their shapes and relatively delicate surfaces. Glass was also cut or engraved to form patterns on the surface or create cameo effects, using cutting materials that were harder than the glass such as metal or diamond. Glazed pottery and glass objects were sometimes painted after firing with pigments of various types, including precious metals. Further decorative glass techniques requiring a subsequent firing stage were discussed in the Shaping section above, such as mosaic glass production and inlay work.

The diversity and overlapping nature of the vitreous silicate materials offers an outstanding amount of information about the process of innovation in ancient societies. Overall, remarkably similar raw materials were used to create quite different products, as different as glazed earthenware and blown glass. At the same time, quite different processes were used to create products almost identical in appearance, as in the different methods of glazing faience. These similarities and differences in production materials and techniques can provide crucial data on the technological aspects of inventing new materials. Understanding the history of development and the distribution of these objects will create insights into the reasons for the development of new materials, whether economic or social, or most likely both. I explore some of these issues for the Indus case in Chapter 6.

METALS: COPPER AND IRON

In this section, I summarize the production processes for copper, from ore or native metal to finished object. I provide a parallel overview for iron and steel, indicating similarities and significant differences in the production processes

Transformative Crafts 145

for these two major metal groups, much as I did for the vitreous silicates in the previous section. Although I have tried to generalize to worldwide patterns, the descriptions here most often match European and Western Asian processing approaches, in large part because of the plentiful historic and prehistoric archaeological and experimental studies for these regions. Craddock (1995) and Tylecote (1987) are excellent general references for anyone interested in metal production, providing detailed summaries of metal production processes for copper, iron, gold, silver, lead, zinc, and other metals, as well as case studies and references to classic works and recent advances. Hodges (1989 [1976]) provides an overview of all the major metals which is useful as a first text due to its brevity and simplicity. However, other references must subsequently be consulted for the great advances in our knowledge of the diversity of ancient metal production practices since Hodges' time, particularly for smelting. Scott (1991; 2002) discusses detailed technical and analytical research on metal production and use worldwide. Henderson (2000) provides updates to Craddock, and additional case studies for Europe and Southeast Asia, while Bisson *et al.* (2000) and Childs and Killick (1993) more fully summarize the African tradition. Piggott (1999a) presents recent overviews of metal production across Asia by regional specialists, from Cyprus to China, and research into the fascinating Mesoamerican and South American metal working tradition has been done by Hosler (1994a; 1994b), Lechtman (1976; 1980; 1988), and Shimada (Shimada and Merkel 1991; Shimada and Griffin 1994), among others. For an excellent summary of the unusual North American case see Martin (1999), as well as recent studies by Ehrhardt (2005), and Anselmi (2004). Specialist publications focused on archaeology and metal production include the journal *Historical Metallurgy* and the newsletter *Institute for Archaeo-Metallurgical Studies (IAMS) News,* both of which frequently publish experimental as well as archaeological research. IAMS, which also publishes monographs, is a major, long-term research center for experimental and analytical work on ancient metal production at the Institute of Archaeology, University College London; a similar center is at the Deutsches Bergbau Museum (German Mining Museum) in Bochum, Germany. Both of these centers have links to research teams working on projects around the world, so their web sites provide a useful entry into current research. MASCA, the Museum Applied Center for Archaeology at the University of Pennsylvania, has produced a number of edited volumes on metals in its Research Papers series that are particularly useful for iron working.

The production of copper and iron objects from ores requires the following steps (Figure 4.11):

1. Collection of native copper or ores (mining)
2. Preliminary processing of native copper or ores (sorting, beneficiation, roasting, etc.)

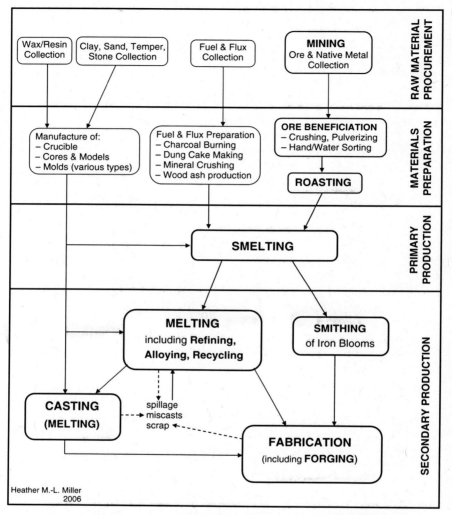

PRODUCTION PROCESS DIAGRAM FOR COPPER AND IRON

FIGURE 4.11   Generalized production process diagram for copper and iron (greatly simplified).

3. Extraction of metal from the ores (smelting) to produce copper or cast iron ingots, or iron bloom
4. Additional melting stages as needed, to purify primary ingots (refining), melt down scrap metal (recycling), or create alloys; solid-state iron refining and alloying (smithing, fining, cementation); note

that of the solid-state refining processes, only smithing is shown in Figure 4.11
5. Creation of metal objects, employing (a) casting and/or (b) fabrication (including forging).

COLLECTION, INCLUDING MINING

Native coppers, which are naturally occurring copper metal deposits, can be directly processed into objects without smelting. As relatively pure copper metal, no melting steps would be necessary to refine this metal, only to either alloy the copper or cast it. In the vast majority of cases, however, objects were made from native copper using primarily hammering, annealing, grinding, and cutting techniques, not casting. In most parts of the world, native coppers would have been depleted rapidly, and copper metal would then have to be extracted from copper ores of various types. The enormous deposits of native copper found in the Great Lakes region of North America are an important exception, where plentiful availability of native copper had a profound effect on the development of metal working, as smelting of copper ores and even melting of copper metal was never a necessity (Martin 1999; Ehrhardt 2005; Craddock 1995). In the North American tradition, metal working is therefore not a transformative craft, but primarily an extractive-reductive one with close technical ties to stone working. This phenomenon creates some very interesting patterns for comparative analysis of technological styles of production once Europeans begin trading with Native North American groups, as discussed in Chapter 5 in the section on Technological Style. Metallic iron, available from meteors, was similarly of significant economic use only in Greenland and northernmost North America, although Craddock (1995: 106–109) notes that analyses of iron objects from these regions also shows a surprising amount of wrought iron (thought previously to be meteoric iron) traded into these areas ahead of direct European and Asian contact.

Elsewhere, people had to develop techniques for extracting copper metal from the ores available. This was also the case for iron, which was only produced in the Eastern Hemisphere (Old World). *Ores* are complex minerals made up of metals, silicates, and other materials, which are heated to high temperatures to melt out the metal desired. Copper ores at or near the surface of the earth are found in the *gossan*, the weathered upper portion of the copper-bearing deposits that have been oxidized. These ores are usually a mixture of native copper, any remaining unweathered (unoxidized) copper sulfide ores, and the copper oxides, carbonates, and other brilliantly colored minerals that were also valued as colored stones and pigments both before and after the development of copper metallurgy. Even if ores are found on a site, small quantities of copper minerals and other metal ores were not necessarily used for metal production, but could have been collected for colorants, face and body painting, or for stone object production, so care has to be taken to ensure that these minerals were actually collected for smelting. The upper weathered deposits also contain iron oxides, which are important components of the copper smelting process, as well as providing the gossan with its red color and alternative name, the "iron hat" (Craddock 1995). However, the deposits of these oxidized copper ores were sometimes quite shallow, requiring exploitation of the secondarily enriched copper ores below the gossen and the unweathered parent ores below this zone. These lower ores are typically sulfidic copper ores, such as chalcocite, chalcopyrite, and other complex minerals containing iron and other metals including lead, antimony, and arsenic. Some of these minerals were also valued for their brilliant colors, as well as being potential sources of other metals besides copper, so that it can sometimes be quite difficult to determine which metals were the focus of a particular mining operation. As has been demonstrated for the Rio Tinto mines in Spain, sometimes the same deposit was worked for different metals during different time periods (Rothenberg and Blanco-Freijeiro 1981; Craddock 1995).

Iron deposits are unusual among the ancient metals in that good quality deposits are ubiquitous, even the higher-grade iron ores needed for early smelting processes. (Note that the iron found in non-ferrous ore deposits as described above would not be ideal for iron production, containing too many trace metals that would make the iron brittle or difficult to work.) Craddock (1995: 235) notes that this widespread availability of smeltable iron ores was quite different from the much more restricted number of deposits of non-ferrous ores, and this great difference in access to raw materials must have had an enormous impact on the organization and control of metal production. The accessibility of iron ore would likely have encouraged experimentation with production of iron objects, even when bronze had superior working qualities and its production properties were much better understood.

The collection of native copper and rich surface ore deposits can be as simple as collecting shellfish or digging for tubers, requiring only baskets and simple wooden digging sticks. Once surface deposits were depleted, shallow pits or deeper trenches would have to be dug to access near-surface deposits of native copper or various ores. Outcrops were also quarried back into shallow caves and tunnels. Some societies also constructed deep mine tunnels, primarily for stone and non-ferrous metal ores, as iron deposits were plentiful enough near the surface that surface stripping was used except for deposits of extraordinary iron ore compounds. While even the most complex mining operations required relatively few tools, much specialized skill and knowledge was needed to deal with dangerous conditions from noxious gases to flooding to tunnel collapse. For shallow or deep mining, tools would be needed to break the ore or native copper out of the surrounding rock and to collect the resulting piles of unprocessed ore or metal. Stone hammers and antler picks were used for ore quarrying and later metal picks and chisels of bronze or iron. For outcrops as well as deeper tunnels, fire-setting might be used, employing methods similar to those used in stone quarrying as described in the ethnographic account of fire-setting in the Stone section of Chapter 3. Fire-setting resulted in a characteristic smooth curving surface of the rock wall, and Craddock (1995) illustrates the marks left on mine walls by different kinds of mining tools, including fire-setting, stone mauls, and metal picks. Based on the few preserved finds, baskets and wooden trays as well as bone and wooden scoops, seem to have been used for collection and transport.

Open pit or trench mining was arduous, although less hazardous than tunneling. These types of mining might require some knowledge of construction and managerial organization, depending on the scale of production. The most important technical knowledge for trench or pit mining, however, would be a good idea of the geological nature of metal deposits in order to follow the deposits most effectively. Trench or open pit work could be accomplished by a relatively wide range of people depending on the intensity and scale of production, from full-time specialists and slaves as described for tunnel mining below, to small groups mining on an occasional or seasonal basis to acquire ore or native metal for personal use or potential trade. For deep mining, whether for metal or stone, additional necessary tools and knowledge included light sources and architectural techniques of safe tunnel construction and water removal. Providing an adequate source of light for extended periods of time was an important issue in mining in the ancient period as well as the last century, as the available light sources (oil lamps, candles, and torches) competed with the miners for consumption of precious oxygen and filled the shafts with smoke. Tunnels of any depth would have required shoring beams, airshafts, and often methods of removing water, as Craddock (1995) describes and illustrates. The more complex mines required a high degree of specialist tunneling knowledge as well as managerial organization and planning. Craddock summarizes the team studies of such mines, including those of prehistoric Wales, Egyptian Timna in the Arabah valley of the Sinai, Roman Rio Tinto in Spain, and early historic Dariba in India. Mining in tunnels was dangerous, exhausting, and poisonous work. As in historic mines of the last few centuries, ancient miners would have had their health degraded and life span shortened by their profession. The same was true for smelting. Managers, and to some extent miners, in large-scale tunneled mines were likely occupational specialists, as expert knowledge would be needed. However, in many of the documented ancient cases, ore extraction in large-scale mining was primarily done by prisoners of war, convicts, or slaves, and was often considered the very worst assignment for a slave. For those cases where we have records, gender and age divisions were also often employed, with men quarrying the ores while women and children removed the ore and processed it.

Processing of Ores and Native Copper; Fuel and Fluxes

After lumps of native copper, copper ore, or iron ore were recovered from deposits, whether by surface collection or deep tunneling, the lumps were usually crushed or pulverized with hammers, grinding stones, mortars, and pestles. In historic periods, water-driven pounding machines (stamp mills) were used in some parts of the world (Craddock 1995: 161). *Crushing* made it easier to separate the native copper metals or the ores from the associated minerals or gangue (usually silicate rocks). *Sorting* was done by hand-sorting of larger lumps, and sometimes water sorting of finer particles, using repeated washing to remove the gangue of lower specific gravity and leave behind the ore, as in panning. Various water sieving apparatuses were utilized in large mining operations, particularly if precious metal was also present, as illustrated in Craddock (1995). For copper and iron ores, an additional benefit of this pounding stage was that the smaller lumps could be more easily smelted. The combination of crushing and sorting to remove unwanted material and provide a higher percentage of rich ore for the smelt is called *beneficiation*.

After extraction from surrounding rock and sorting, native copper metal could be directly fabricated into objects and sheets. The pulverized copper ores usually underwent a *roasting* stage, heated either in an open fire or in an oven with an ample air supply, with much raking and turning of the ores. This helped to remove any copper sulfides present by converting them to oxides; some copper sulfides were often present even if primarily oxide ores were used. Roasting also converted the iron sulfides within the copper ores into iron oxide. The iron oxide was removed by combination with any remaining silicates from the copper ores (or by adding a silicate flux if needed), to form an iron-silicate slag. On cooling, this slag could be removed from the roasted ores, giving a better product for the smelting stage. Such an iron-silicate slag also creates considerable confusion among unwary archaeologists who find them, as to whether these slags represent the remains of copper or iron production. For cases where primarily copper sulfide ores were being processed, this roasting stage might not be enough to convert all the copper sulfides to oxides. Instead, the roasted ores were then smelted to produce a copper-sulfide *matte*, which would then be crushed and roasted again to convert the matte to copper oxide. The copper oxide would then be smelted to copper metal as usual. Craddock (1995: 149–153) notes that there were frequently multiple cycles of roasting and smelting employed for copper sulfide ores in post-medieval Europe and historic Asia before the final smelting to metal could take place.

Fuel was also an important material to be collected and processed for the smelting, melting, and forging stages, for both copper and iron. Both copper and iron smelting and working need reducing conditions (little oxygen), and one of the best ways to achieve this is by using fuels that scavenge oxygen when they burn. Such fuels are particularly useful when the metal craftsperson needs access to the metal in stages like forging and so cannot restrict completely the flow of air to the metal. Metal production also requires relatively high temperatures. Charcoal, specially prepared wood, is a good fuel to achieve high temperatures, to provide a strong reducing atmosphere, and to avoid the introduction of negative elements such as sulfur and phosphorus (Horne 1982; Tylecote 1987; Craddock 1995). Furthermore, for copper melting and fabrication, use of charcoal helps to prevent the formation of copper oxides, which inhibit melting and form scale on objects during annealing (McCreight 1982, 1986). Particular types of charcoal are usually preferred, from hot, fast-burning hardwoods; as with pottery kilns, paleoethnobotanical analysis of the charcoal from smelting sites, including fragments found in slags, can identify the particular woods used. It is also possible that dung fuel could have been employed for specific needs in the past, especially smithing and annealing stages that require regular heat over a long period of time. While dung fuels produce a very long-lasting, steady heat, they do not typically produce a rapid, high heat, and so their often phosphorus-rich ash would be detrimental for smelting and melting. These same characteristics might be advantageous in iron smithing, however. Paleoethnobotanical analysis could also help to distinguish the use of dung, through identification of characteristic assemblages of seed species typically found in animal dung (Charles 1998).

Coal is also an excellent fuel for metal production, even better than charcoal for smelting since it is stronger and so able to support the weight of the prepared ores in a furnace without collapsing and extinguishing the smelt (Craddock 1995). However, the majority of the world's coal contains sulfur, which affects the smelt and contaminate the metals, and so generally cannot be used without further processing to remove sulfur and produce *coke*. The production of coke occurred only within the past 300 years in Europe, but coke production as well as other methods of using sulfur-rich coal effectively were developed at least a thousand years earlier in China. This widespread use of coal in China so much earlier than in Europe is probably due to the relatively high availability of anthracite coal (coal without sulfur) in China, encouraging early coal use so that methods were subsequently sought of using the other sulfur-containing coals (Craddock 1995: 196; Hodges 1989 [1976]: 89;

Bronson 1999). The lack of timber in some areas of China may also have played a role. The deforestation of areas near large-scale smelting operations in various parts of the world is well-attested in the archaeological and historical record, as one of many environmental effects of metal production, so that even timber-rich areas like northern Europe were eventually encouraged to find alternative fuel sources in the form of processed coal.

Finally, *fluxes* were needed at various points in the process of metal production. The term "flux" is used for two distinct types of materials, one set used during smelting and sometimes melting, and a different group used for hot-joining during fabrication. Fluxes employed during smelting are minerals added to the smelting charge to lower the fusion temperature, as described below. These fluxes supply the appropriate elements/compounds to remove *gangue*, the unwanted nonmetal portions of the metal ores. Many of these fluxes were the same as those used as glass modifiers, such as soda, potash, or metal oxides; for example, the use of iron oxide as a flux for roasting and matte production from copper sulfide ores was described above. As another example, the blast furnace method of producing liquid iron required the addition of lime or calcium-rich clays to remove silica and other gangue materials, since the usual iron-silica slags would be reduced to metal (Craddock 1995: 250). Fluxes employed in fabrication were various materials used in joining by soldering (see below), to aid the flow of the solder and to protect the metal parts being joined from oxidation. These are different materials from the smelting and melting fluxes, and are also specific to the particular metals used, although borax mixed with other materials is a common soldering flux.

SMELTING

The next stage in metal production is the extraction of metal from the ores, *smelting*, to produce primary ingots. The usual system for smelting, at least in ancient Europe and Western Asia, was to heat the pulverized ore with charcoal in firing structures of varying types, after the structure had been preheated. For copper production, the aim was to properly heat this mixture (the *charge*) so that the copper minerals would be reduced to copper metal, with the liquified metal flowing to the bottom of the firing structure and the less dense silicate slag floating on the surface of the liquid metal. Fluxes, which could be present in the copper ore, the fuel ash, or the firing structure walls, or which were often intentionally added, would convert any siliceous material present in the ore to glassy slag (*scoria*). For iron, a similar situation would be desired for the production of liquid (cast) iron, as was produced in China using the indirect blast furnace method from the first millennium BC and much later in other parts of the world (Craddock 1995). Outside of China, the product of an iron smelt was a *bloom*, a solid mass of iron metal containing bits of slag and other materials, which was produced at considerably lower temperatures than liquid iron. The metal produced by this direct bloomery process was consolidated and nonmetal inclusions were removed by *smithing* (repeated hammering and annealing) to produce wrought iron, as discussed in the Refining and Alloying section below.

The main challenges for most early smelters were the complete reduction of the mineral to metal and its full separation from the siliceous material and/or the slag. For this to occur, the temperatures had to be high enough, the atmosphere had to be sufficiently reducing (no oxygen), and as much of the siliceous material as possible should be separated from the metal product. In copper and liquid iron production, this separation was achieved by the liquid metal passing through the charge to the bottom of the furnace, and the siliceous materials changing to slag and rising to the top of the mixture. Bloomery iron production operated somewhat differently, of course, as the iron was not molten and not separated out from the slag in entirely the same way, but conversion of the siliceous materials to slag and removal of as much of the slag as possible was also a major goal in bloomery iron production.

To raise temperatures, metal workers developed different ways of increasing airflow and also retaining heat. This included both various firing structure (furnace) designs and various methods of providing a draft, including blowpipes, bellows, and natural draft. Numerous experimental and ethnoarchaeological studies have examined these issues for both copper and iron production, as summarized and referenced in Craddock (1995) and Henderson (2000). Retention of heat and flow of air were two aspects of furnace design that had to be balanced, along with other attributes. For example, covered firing structures had to be at least partially destroyed after every smelt, but the covering retained heat better than open firing structures, allowing reduced fuel use and a faster smelting time. Some firing structures, such as most shaft furnaces, could not only reach higher temperatures through increased draft, but could also be recharged to some extent during the smelt, allowing for a larger amount of metal in the final ingot. Achieving temperature and reducing conditions sufficient to create a liquid slag aids greatly in removal of the slag from the metal, and *tapping* the liquid slag, removing it from the firing structure during the smelting process through a tapping hole, allows for a longer smelt with a greater quantity of metal produced, as the slag does not choke the structure. Craddock (1995: 174–189), in his description of a range of draft production methods, stresses that the most important quality in draft production is that the air supply be steady and controlled. He has some doubts about the operation of most wind-blown furnaces on this account, with the exception of the Sri Lankan iron smelting furnaces investigated by Juleff (1998), and possibly the Sub-Saharan African updraft iron smelting furnaces. At the same time that an increased draft might be used to raise the temperature, the metal worker had to be sure to maintain a reducing atmosphere. To ensure a reducing atmosphere through the ubiquitous presence of carbon, various methods were used of placing the ore and charcoal in the furnace to ensure free circulation of gases: as a uniform mixture, or in alternating layers, or even formed into balls, using dung as an adhesive and likely as an additional fuel and reducing agent (Craddock 1995). Finally, the type of ores and fluxes used could raise or lower the temperatures needed to produce molten metal. The conversion of siliceous mineral matter to liquid metallurgical slag was expedited with the addition of fluxes if needed, as noted above. It is clear from these examples that the particular conditions of each smelt depended on a variety of highly interconnected factors.

The process of smelting in furnaces described above is the most common method for copper smelting, but Craddock examines in detail the archaeological, experimental, and theoretical evidence for direct copper *crucible smelting*, which now appears to be the earliest method of copper smelting in Eurasia. The investigation of crucible smelting of copper is a relatively recent phenomenon, although Tylecote (1974) suggested some decades ago that it might be possible to smelt in crucibles. Craddock (1995: 126–143) outlines evidence from several early sites, particularly Feinan in Jordan and sites on the Iberian peninsula, to show that the more familiar slag-tapping furnaces are a later development and the earliest copper smelting took place in crucibles or large open-bowl furnaces heated from above and with little or no slag production. All types of copper ores appear to have been used, including direct smelting of sulfides. Pigott (1999b) discusses additional examples and experimental studies focused on direct crucible "co-smelting" of copper oxide and sulfide ores. The copper metal produced in slag-tapping furnaces has a much high iron content than the crucible-smelted coppers, even after refinement through remelting. Craddock uses this difference in iron content as a clue to trace the different development of copper smelting in Western Asia and the eastern Mediterranean compared to western Europe, using the average iron content of assemblages (not individual items) from these regions between the early third millennium and the first millennium BCE. His results indicate that crucible smelting was still used to the west, except for specific locations, while the eastern regions of the Mediterranean and Western Asia had switched primarily to slag-tapping furnaces.

Information about metal production can be gained from the study of the non-metal by-products of metal processing, as well as metal products. Products, such as metal ingots and objects, are occasionally found at production sites, but these metal pieces are unlikely to be discarded in any quantity since failed metal products can easily be remelted and recycled. Therefore, "wasters" of misfired metal do not accumulate at metal production sites, unlike the failed product wasters characteristic of pottery production sites. Even if lost in antiquity, the metal products are highly vulnerable to decay or collection by later people for recycling. Therefore, like the debitage from stone working, by-products from metal production provide essential information about the method and efficiency of processing, the technologies involved, the temperatures reached, and the types of fuel used, as well as information about metal composition (e.g., Bachmann 1982; Bayley 1985, 1989; Craddock 1989, 1995; Cooke and Nielsen 1978; Freestone 1989; Tite, et al. 1985; Tylecote 1980, 1987). These by-products comprise a diverse group of discarded raw materials, tools, and processing residues, many of which are highly weather-resistant and archaeologically visible (Figure 4.12). Smelting, other than crucible smelting of relatively high-quality ores, leaves large amounts of weather-resistant metallurgical slags, vitrified masses of silica and other fused minerals that generally accumulate in conspicuous mounds near the smelting furnaces. Such metallurgical slags (scoria) are seldom collected and used for any purpose (although there are some exceptions). These slag heaps make it easy to identify the location of past smelting activities, even if they are usually difficult to date. In contrast to smelting, melting for refining or recycling usually leaves little if any metallurgical slag, only fragments of crucibles, furnace linings, and possibly molds for secondary ingots, in relatively small quantities. As noted,

Material Type	Smelting (with Slagging)	Melting of Metal (& Crucible Smelting!)	Non-metallurgical Production (pottery, glass, etc.)
Ore/Flux	Fragments of ore/flux usually found with slags Proximity to ore source	Fragments of ore/flux rare/none Proximity to markets	No associated ore/flux
Firing Structures (furnaces, hearths, kilns)	No Ash Diameter usually <60 cm Heavily vitrified Usually poorly preserved (destroyed to remove smelt)	Ash possible Small or non-existent Vitrified or not - variable Usually less poorly preserved (not destroyed to remove melt or smelt)	Ash possible Diameter may be large (>60 cm possible) Tend to be unvitrified, but may be ash-glazed Usually better preserved
Kiln tools/furniture (crucibles, molds, tuyeres, etc.)	Heavily vitrified; Crucibles and molds unlikely	Some vitrification or ash-glazing; Crucibles and a variety of mold types possible	(Different types of kiln tools/furniture)
Other vitrified materials	Large quantities (many kgs) of hard, dense scoria, dark in color with relatively uniform structure and fewer, larger bubbles (includes both furnace bottoms and tap slags)	Slags much more vesicular/porous, lighter weight (usually vitrified crucibles, furnaces, etc.); less homogeneous, inclusions distributed heterogeneously; macroscopic metal inclusions possible/likely	Usually much lighter in color and density, but not always; also very unhomogeneous.

FIGURE 4.12 Typical assemblage characteristics for non-ferrous metal processing. (Compiled from Craddock 1989:193, Fig. 8.2; Bayley 1985; Cooke and Nielsen 1978.)

metal products and scrap are likely to be recycled not discarded (Figure 4.11), so that only rarely are forgotten hoards of scrap metal discovered.

REFINING AND ALLOYING

Smelting normally takes place near to the mining sites, especially for non-ferrous metals with many fewer sources of ore. Primary ingots or native copper lumps are then usually taken to habitation places, and sometimes traded great distances by land or water. Whether traded or kept by the people who recovered this metal, the ingots or lumps can be stored for considerable periods of time until needed for use or trade. A major issue in archaeological studies of ancient copper production and distribution has been the measurement of the impurities in copper metal objects and ingots in order to locate their original ore source (*sourcing* or *proveniencing*). Some of the most extensive of such studies have focused on the trade in copper and its alloying materials across ancient Europe, the Mediterranean and Western Asia; Henderson (2000: 248–261) summarizes much of this research for both chemical composition and lead isotope studies. The process of metal sourcing is complicated by a wide range of difficulties, from the addition or loss of particular trace elements during processing to the mixing of metals from different sources during recycling of scrap metal.

Primary ingots from smelting typically required further processing before they could be used for metal production, as did iron blooms. Such *refining* processes might take place near the mining and smelting sites, especially for large-scale mining and smelting operations, or might take place at consumption sites nearby or a considerable distance away. Iron bloom smithing, and melting of primary ingots to create secondary or refined ingots, are both undertaken to remove slags and other undesired elements left in the original smelting blooms and ingots, and sometimes to break up large smelting ingots into more workable or transportable ingots (Figure 4.11). *Smithing* is the process of repeated cycles of heating the bloom to red heat to melt slag particles, hammering to squeeze out the slag, and annealing again, eventually producing wrought iron. "Cast" (liquid) iron, was frequently very brittle due to high carbon content, but could be refined by *fining* (not shown in Figure 4.11). In fining, a blast of air was passed over the surface of the heated iron fragments, burning out the carbon and other impurities as the metal was raked and turned in a semi-molten state. The metal produced from fining was in the form of a bloom, and was then forged as usual (Craddock 1995: 253; Hodges 1989 [1976]: 90). Finally, while primary copper ingots could be used directly for fabrication, they were likely re-melted in most cases to *refine* the primary ingots, which often still contained undesirable impurities such as iron, silicates, sulfur, arsenic, or other metals. Care had to be taken to avoid too much exposure to oxygen, which would create copper oxides. This could be done by using crucibles that exposed only a small surface area of the molten copper, or by stirring the molten metal with green branches or twigs (*poling*). Melting of copper was a good way to remove sulfur and arsenic as gases, and iron and silicates could be removed as slags by the addition of a small amount of additional flux such as clean sand, if needed (Craddock 1995: 203–204; Hodges 1989 [1976]: 69–70).

The production of alloys can take place at any one of a number of stages during the production process. Alloy production usually involves melting, but can sometimes employ solid-state processes, as in iron carburization. One problem in discussing alloying is determining what constitutes an alloy and what a single metal with impurities, as different researchers have used different standards to define alloying. "Intentional" alloys can be defined as more than 1% of an element, or more than 2%, or more than 5%. Stech (1999) provides a thorough discussion of the problem of alloy determination, and advises that in these lower percentages (less than 5%), it is often not possible to determine if the "alloy" is the result of the intentional mixture of two separate metals or metal ores, or due to the natural metallic impurities in particular ores, not to mention the re-melting of a mixture of metal objects in scrap recycling. For example, in many cases tin is unequivocally an intentional alloy with copper, while an equal level of lead, often found as an impurity in copper ores, might or might not be the result of intentional alloying. Possible patterns of alloying are also obscured archaeologically by the lack of a large sample, problems with chronological control, or the inconsistent manner in which samples from different sites have been studied.

Patterns of alloying and reasons for particular alloys vary from society to society, as is striking in the emphasis on copper alloys made with gold and silver in Central and South America, in contrast to the focus on copper alloys made with tin, lead, and perhaps arsenic (deliberately added or not) in Europe and Western Asia. Color, sound, or the ability to resist oxidation may have been more important than hardness or strength for particular purposes or societies, as is the case for historic Mexico and South Asia (Hosler 1994a, 1994b; Lahiri 1995, 1993; Chakrabarti and Lahiri 1996; Craddock 1995: 285-292 also summarizes research by several teams on similar issues for arsenical copper in Europe). Alloying can be used for a variety of purposes: functional, aesthetic, ritual, or simply expedient. The addition of tin to copper may have been done to increase strength and hardness for some objects, but may have been used to produce particular colors or fulfill ritual requirements in other objects. Some ancient metalsmiths may not have followed a rigid system of alloying related to specific artifact categories, or a mixture of alloyed scrap metals may have been the material available for a smith's selection—expediency is difficult to model archaeologically, but too common ethnographically to ignore. When faced with the choice of desired characteristics, including hardness and color, ancient metalsmiths may have chosen between a number of alternative means of producing a given result. For example, in some instances they may have relied on physical modifications such as forging to harden metal, while in other situations they may have chosen to produce a harder metal by modifying the composition of the metal through alloying. These choices would depend on the manufacturing techniques used, the types of metal and alloys available, and the stage of metal production (smelting, melting, casting of blanks, etc.) at which the end product was first visualized.

For iron, the primary alloying element is carbon. Cast iron is iron containing 2 to 5% carbon, which lowers the temperature enough that it can be melted and cast, but which also creates brittleness. Very early cast iron production is documented for China, much earlier than the rest of the world. Steel, the other main iron alloy, contains less than 1 or (at most) 2% carbon, and is much more malleable and strong under proper heating and cooling conditions. Steel can be created either by remelting, as in crucible steel production, or by solid-state methods of fusion in cementation or carburization. *Crucible steel production* was used to create the famous *wootz* steel used to make damascus blades and other steel objects, first in South Asia and subsequently in Western Asia as well. Wrought iron was placed in small crucibles with organic materials and heated to very high temperatures for a long period of time. Once the iron absorbed enough carbon from the organic materials, the melting point of this incipient steel would be low enough for it to melt, forming a homogenous steel ingot without slag inclusions. Craddock (1995: 275–283) discusses and updates the previous literature summary on crucible steel production by Bronson (1986), including recent work by Lowe (1989a; 1989b) and Juleff (1998), and provides a number of outstanding illustrations of crucibles, lids, and ingots from crucible steel production sites in South Asia. *Cementation* or *carburization* produced steel from iron without melting, in a process physically similar to the cementation glazing method described for faience in the Vitreous Silicates section. Iron fragments were placed together with powdered carbon or organic materials in closed containers, and heated for long periods at high temperatures but below the melting point of the iron. The iron absorbed the carbon, creating steel.

Metals are also melted at this stage of production to recycle scrap metal (Figure 4.11). Scrap melting is a very under-represented industry in the archaeological record, as are melting and fabrication stages in general, but was probably one of the primary methods of metal acquisition. It is likely to have taken place near to consumption areas from which the scrap was collected. All of the processes of melting, whether for refining, alloying, or recycling, resulted in the production either of secondary ingots, casting blanks, or sometimes even direct casting of finished objects.

Shaping and Finishing Methods: Casting and Fabrication (Including Forging)

Shaping techniques for metal objects can be classified depending on the state of the metal during working. *Casting* refers to the manipulation of molten metal, while *fabrication* is the treatment of non-molten metal, whether cold or hot. These two categories divide the methodologies of metal working artisans as well as the states of the metal itself. Fabrication involves the direct shaping of metal, while casting begins with the shaping of other materials into which the molten metal is poured. The tools and techniques of the two categories overlap to some degree, and ancient metal working ateliers may have been involved in both fabrication and casting. Some objects, however, may have been cast by one group of artisans and finished or fabricated by another group in a separate workshop. The possible division of manufacturing stages into discrete and often exclusive activities practiced by different artisans is an important part of metal working that has not been well investigated for most regions, primarily because few metal working areas (as opposed to smelting and refining areas) have been conclusively identified. Instead, much of the evidence for casting and fabrication techniques comes from the examination of finished objects. Casting includes open, bivalve, and multi-piece casting, as well as lost wax or lost model techniques for non-ferrous metals. Fabrication techniques include shaping by forging and annealing to manufacture sheets, vessels, and other objects, as well as cutting, cold and hot joining, and decorative finishing methods such as polishing, engraving and inlay.

Casting

Melting or re-melting of metal is also a necessary stage in the production of cast objects, both semi-finished and finished, in order to pour the molten metal into molds of various types. The best evidence for metal casting activities at a site is the presence of molds. Ancient mold types include open stone, metal, terracotta, or sand molds; bivalve and multipiece stone, metal, terracotta, or sand molds; and "lost-model" molds of sandy clay. Horne (1990) suggests the term "lost model" rather than "lost wax" since the technique employs other materials besides wax, such as tallow, resin and tar. Sand casting and lost-model casting both leave almost no archaeological traces. Sand-based molds are used for casting both ferrous and non-ferrous metals, employing a finely powdered sand that is usually mixed with water and organics such as

FIGURE 4.13 Pouring molten copper alloy into bivalve sand mold. *Photo courtesy of Lisa Ferin.*

dissolved sugars to act as an adhesive (Mukherjee 1978; Untracht 1975). This mixture can be used to make an open mold or packed into a hinged wooden box to make a bivalve mold (Figure 4.13 and 4.14). A form made of wood or some other material is impressed into the sand mixture, which is cohesive enough to create a mold that can even be set on edge to pour in the molten

FIGURE 4.14 Opening bivalve sand mold to reveal copper alloy bowl. *Photo courtesy of Lisa Ferin.*

metal. This method works well for the production of flat objects, such as blade tools, but is also commonly used for three-dimensional objects, as shown in the casting of a high-tin copper bowl in Figure 4.14. It may even leave characteristic flashing lines on the objects, often taken to be indicative of the use of stone or terracotta bivalve molds. Since forms are used to impress the sand and create the mold, some degree of duplication of objects is possible, and creation of the sand molds is obviously quite rapid. Although these molds have a great resistance to heat, making them an excellent casting material, they break down quickly into sandy deposits when exposed to weathering from water and wind. In addition, modern sand molds are usually crushed and reused, and ancient molds would probably have been similarly recycled.

The materials used for lost-model molds are also quite ephemeral. Forming a continuum with the fine sticky sand used for sand casting, lost-model molds employ a more cohesive sandy clay so as to better retain the complex three-dimensional features of the object to be cast. Several grades of material are often used. The model of wax, resin or tar is first coated with a very fine sandy clay. This inner coat will form the details of the object to be cast, so the finer detail desired, the finer the texture of the coat. Increasingly coarse sandy clay is used to form the bulk of the mold, allowing the permeation of gases through the very fabric of the mold. Organic materials are often mixed with the clay coatings for strength and perhaps to provide a more reducing atmosphere. The crucible containing the metal can be built onto the mold, as is done in sub-Saharan Africa and parts of India, or metal can be poured into the mold from a separate crucible, as was done in the Americas, Egypt and India (Emmerich 1965; Fox 1988; Fröhlich 1979; Horne 1987, 1990; Mukherjee 1978; Reeves 1962; Scheel 1989). An essential component of the lost-model process is the use of a sandy clay that will not sinter under high temperatures, as this would hinder the escape of gases and encourage the cracking of the mold during casting. Thus, in addition to the fact that the molds are broken to remove the cast object, such molds also break down very quickly when exposed to weathering. As with sand casting, the broken pieces of the mold are also often recycled in ethnographic cases, increasing the likelihood of their archaeological invisibility. A major advantage of the use of sand and lost-model molds for craftspeople is the widespread availability of these materials locally, although stone or clay molds could be carried by traveling metalworkers who might not know if proper sands would be available in new regions. The ability to rapidly produce molds and to recycle the mold materials is another of the great advantages of sand and lost-model casting over stone mold casting. While this is beneficial for the artisan, it is a nightmare for the archaeologist. Perhaps with increasing awareness of these methods and their ephemeral remains, archaeologists will begin to look more closely at patches of sand for the tiny fragments that may still retain the contours of cast objects.

Fabrication

Fabrication of metal objects includes all of the various types of modification of non-molten metal: *shaping*, via forging, turning and drawing; *cutting*; cold and hot *joining*; and *finishing* via planishing, filing, polishing, coloring, engraving, and so forth. The metal can be worked while cold or hot, but at a heat below the molten state. Fabrication can include numerous intermediate stages and semi-finished products. Ingots or cast semi-finished objects can be worked directly into a finished object, or ingots can be first forged into sheet form, and then made into objects using various techniques (Tylecote 1987; Hodges 1989 [1976]). Particularly useful publications for examining the wide range of fabrication techniques from the perspective of metalworkers themselves are Untracht (1975), Ogden (1992), Bealer (1976), and Mukherjee (1978), the last an ethnographic survey of modern Indian metal fabrication techniques that includes drawings of products, tools and firing structures.

Shaping

In its broad sense, *shaping* is the controlled mechanical stretching of metal. This includes stretching by forging, including sinking and raising; by spinning or turning; and by drawing. The most common form of shaping is *forging*, "the controlled shaping of metal by the force of a hammer," usually on an anvil or stake (Figure 4.15) (McCreight 1982: 36). Although the term "hammering" is sometimes used for non-ferrous metals, and "forging" reserved for ferrous metals, forging is the term most often used by coppersmiths themselves, and will be used here for all metals. The hammer and the anvil or stake can be made of a variety of materials, such as metal, stone, wood, bone or horn, or even leather; hematite and magnetite nodules may have been particularly valued as hammers prior to iron production. Forging sites can be very ephemeral (Figure 4.15), so finds of such tools, or of the marks left on objects by such tools, comprise one common type of archaeological evidence for forging. The main source of evidence for forging comes from metallographic examination of artifacts. Forging can be done while the metal is hot or cold. *Annealing* is the reheating of an object after working, allowing continued forging without tearing the metal or creating an overly brittle object. Forging not only shapes the object, it also hardens it, and so forging is an important step in the manufacture of edged tools. Thus, most metal working, especially iron working, involves cycles of annealing and hot or cold "hammering" (forging). (See Tylecote (1987), Bealer (1976), and Craddock

FIGURE 4.15   Blacksmith forging iron bar on expedient "anvil" in front of annealing fire.

(1995: 237) for details of iron and steel forging.) *Sheet manufacture* is a type of forging, and is particularly used for non-ferrous metals. *Sinking* and *raising* to form vessels from metal are also types of forging (Figure 4.16). As the names imply, sinking is the forming of metal by hammering from the interior of an object into a depression in an anvil, while raising employs hammering from the exterior of the object over a shaped stake or form. There are a number of ways to raise objects: both from sheets and directly from ingots, while the metal is cold and while it is hot, by an individual artisan or by a group working together. *Spinning* and *turning* are methods of mechanical stretching with results similar to sinking and raising but using a lathe rather than a hammer. Wire production can take place by forging or by *drawing*, pulling the wire through successively smaller holes in a drawplate (Hodges 1989 [1976]: 76; Tylecote 1987: 269–271).

Cutting

Many non-ferrous thin objects were cut out of sheet metal. Cutting was likely done for the most part with chisels. Worldwide, a very common procedure for

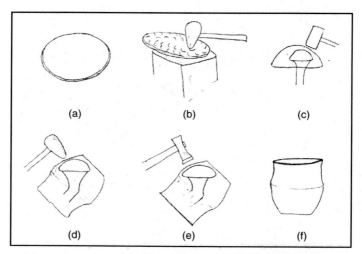

FIGURE 4.16 Sinking and raising vessels from flat metal disks or sheets (a) flat metal disk; (b) sinking disk into wooden block with metal hammer; (c) bouging vessel over stake with wooden mallet; (d) raising vessel over stake with metal hammer; (e) planishing vessel over stake with metal hammer; (f) finished vessel. (Drawn after McCreight 1982.)

both thick and thin non-ferrous objects seems to have been to cut a groove in the metal mass on one or more sides, then snap the piece in two, as was described for stone working in Chapter 3. This groove-and-snap method was largely replaced by sawing in the Eastern Hemisphere after the widespread availability of iron and steel blade and wire saws, chisels, and other cutting tools. For iron working, chisels and punches were used to cut either hot or cold metal. Elaborate "cutout" shapes were made from prepared native copper sheet by the Hopewell people of North America by first incising the outline of the shape on one side of the sheet, then grinding the raised projections on the other side to release the shape (Martin 1999; Ehrhardt 2005; Anselmi 2004; Craddock 1989).

Joining

Hodges (1989 [1976]: 76–77, 86–87) provides a good overview discussion of joining methods. Cold joining is the joining of metal without heat, and is largely used for non-ferrous metals. It primarily involves the use of *rivets* or similar pins to attach pieces of metal together, such as securing metal handles to metal vessels. The more unusual practice of solid phase welding, applying pressure using vigorous burnishing and annealing, can also be classified as a form of cold joining which was used for gold and other very soft metals. In hot joining, the body of the metal is not molten, but the joining material can be molten. For non-ferrous metals, *soldering* (pronounced "saudering") is the most common method of hot joining, where a metal alloy that melts at a lower temperature than the metals to be joined is applied as the solder, a fluxing material is used to protect the metals, and the entire object is heated to melt the solder. A similar method, brazing, was used for ferrous materials. "Running-on" is the rather inelegant but effective hot joining method of pouring a small amount of molten metal over a join, and "casting-on" is a similar but more intricate addition of a cast piece onto an existing metal object. For iron, the most common method of joining is *welding*, joining two pieces of iron by hammering them together when white-hot.

Decorative Finishing

Major decorative finishing techniques include the planishing of forged (especially raised) objects; polishing and filing to smooth surfaces; engraving; surface coloration via plating or enrichment; and inlay. The most common decorative forging method is *planishing*, fine, even hammering with a highly polished hammer to create a smooth, even surface particularly on forged or raised objects (Figure 4.16). Other decorative forging techniques include the use of stamps or punches; hammering of thin metal sheet, often gold, into or over patterns; and chasing, the working of metal from both surfaces. *Polishing* techniques are the same as those described for stone objects in Chapter 3, and polishing materials used for metal objects include metal files, ground and polished stone hones, sand- or silt-sized powders, wood or siliceous plant parts, or even leather. *Engraving*, usually with a metal point, is used to produce designs or accentuate details by carving into the metal surface. *Surface coloration* can be done by coating with another metal, as was done in tinning; by chemical enrichment of the surface, through plating or gilding; by chemical depletion of the surface, as in pickling and depletion gilding; or by the many chemical changes involved in patination (Brannt 1919). Pattern welding is a decorative method used for iron blades employing surface coloration through the carbon enrichment of the blade surface and subsequent bending, and is best known as the method used to make the distinctive patterns of iron and steel in the production of Japanese Samurai swords (Craddock 1995: 271–275; Hodges 1989 [1976]: 88). Finally, *inlay* work includes the inlay of other metals and stone, using a variety of setting methods, including hammering, melting, and cold joining. Many of these and other decorative finishing methods for non-ferrous materials are discussed further in Untracht (1975), Ogden (1992), and Brannt (1919).

This introductory background to the processes of production for the major craft groups presented in Chapters 3 and 4 has already provided some insights into differences and similarities between these crafts and their practice. The next two chapters examine past technologies and archaeological approaches to technology from a different perspective: that of thematic investigations of entire technological systems as part of social needs and desires. Chapter 5 begins with a comparison of the reed boat technologies of coastal Southern California and the Arabian Sea, at the same time introducing most of the approaches employed by archaeologists to understand technological systems in the past.

CHAPTER 5

Thematic Studies in Technology

> At first I could see nothing, the hot air escaping from the chamber causing the candle flame to flicker, but presently, as my eyes grew accustomed to the light, details of the room within emerged slowly from the mist, strange animals, statues, and gold—everywhere the glint of gold. For the moment—an eternity it must have seemed to the others standing by—I was struck dumb with amazement, and when Lord Carnarvon, unable to stand the suspense any longer, inquired anxiously, "Can you see anything?" it was all I could do to get out the words, "Yes, wonderful things."
> (Carter and Mace 1923: 95–96)

Howard Carter's first view of Tutankhamen's tomb, through a candlelit peephole, is often used to illustrate the marvelous discoveries made by archaeologists, discoveries of tombs and treasures. The thematic studies in the next two chapters illustrate the "wonderful things" revealed by the archaeological investigation of past technologies. I can only pick out a shape here and a gleam there among the vast amounts of research into ancient technology by archaeologists and other researchers of the past. Some of the gleams are precious objects, famous studies and classics in the field. Some of the shapes are new investigations, still blurred at the edges, but intriguing in their possibilities.

In fact, the actual practice of archaeological investigation is better illustrated by the subsequent sentence in Carter's account, the seldom-quoted final sentence in the chapter describing the finding of the tomb.

> Then widening the hole a little further, so that we both could see, we inserted an electric torch.

While not as romantically appealing as the candlelit glimmers of "wonderful things," the use of electric torches was extremely important in the proper documentation of this tomb and its subsequent analysis, particularly in allowing painstaking photographic documentation. Such a use of "cutting-edge" scientific tools to (literally) shed light on archaeological questions is an even more fitting metaphor for the archaeological study of technology. Most importantly, it was the possibility of carefully examining and documenting the collection as a whole, still in context, that made the Tutankhamen tomb so important. It is the total technological system, as glimpsed through many different objects in context, which truly reveals the lives and patterns of the past to us. Therefore, in the next two chapters I examine five general themes through a number of studies from around the world, to show how investigations of ancient technology can provide information about a range of topics.

I return, then, to my original questions: how and why do archaeologists study the technologies of past peoples, and how does this help us to understand past societies? I addressed these questions by definition in the first two chapters, and now I will address them by illustration, using first the example of reed-bundle boat technology in the Arabian Sea and Southern California. This first topic shows how archaeological finds, laboratory analyses, ethnography, and experimental studies are employed to provide information not only on production techniques and processes, but also on the role of these boats in the economic systems of very different societies.

TECHNOLOGICAL SYSTEMS: REED BOAT PRODUCTION AND USE

> Again, the ship and the tools employed in its production symbolize a whole economic and social system.
>
> (Childe 1981 [1956]: 31)

A brief overview of boat technologies from two different parts of the world and two different time periods provide instructive examples of archaeological investigations of past technologies. These examples also illustrate *why* archaeologists study ancient technology—for the information it gives us about the development and acceptance of new objects and new production techniques, and about changes in past economies, social structures, and political organizations. These two examples show how boat construction and use were part of a technological system impacting the mechanisms of exchange, wealth accumulation, and economic-based power in these two societies.

Reconstructing Reed Boats and Exchange Networks in the Arabian Sea

Some eight thousand years ago, as early as the sixth millennium BCE, direct evidence is available for the types of watercraft used by people in Mesopotamia and the Arabian Sea (Crawford 2001; Schwartz and Hollander 2006; Cleuziou and Tosi 1994; Vosmer 2000). Among these craft were true boats, having the ability to displace water, rather than rafts whose buoyancy relied on the buoyancy of the construction materials themselves, in this case reeds (Johnstone 1980; McGrail 1985). The boats themselves have not been preserved, but pieces of the tar-like coating of bitumen over their exterior surface have been found at several archaeological sites. Both wooden and reed watercraft were used in the rivers of Mesopotamia as well as the marine waters of the Persian Gulf and Arabian Sea (Figure 5.1). The majority of the analyzed bitumen fragments from boats come from the fourth and third millennia BCE, especially from the site of Ra's al-Junayz in modern-day Oman. In several buildings dating to about 2500–2200 BCE, archaeologists found more than 300 pieces of bitumen, a natural petroleum tar-like substance, which were used to waterproof watercraft made of reed bundles as well as wooden

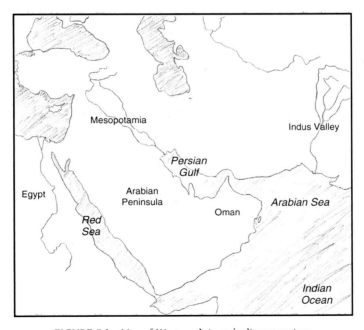

FIGURE 5.1 Map of Western Asia and adjacent regions.

plank boats (Cleuziou and Tosi 1994). For the reed watercraft, impressions of ropes and reed bundles in these bitumen fragments have allowed archaeologists to reconstruct the boat shapes and construction methods. The presence of barnacles on the exterior of some of the bitumen fragments verified that this bitumen had indeed been used on seagoing boats.

Additional evidence specific to the construction and use of these boats comes from ancient drawings and models of reed boats from this region, and from ancient texts. This information has been expanded and clarified using general principles of naval engineering (Vosmer 2000), as well as descriptions and examples of reed boats used in recent and historic times. The most useful ethnographic evidence for construction of these reed boats has come from the modern and historic Middle East (Heyerdahl 1980; Thesiger 1964; Ochsenschlager 1992). With the similarity of available resources and environmental conditions in the recent Middle East and ancient Mesopotamia and Arabia, such ethnographic cases are likely to be the most useful for ancient reconstructions. But reed-boat building techniques used in other parts of the world have also been investigated, to help in the assessment of possible changes in techniques over the past thousands of years. Such studies of the actions of living people to understand the clues left by the actions of past people is the basis of ethnoarchaeology (David and Kramer 2001). Where possible, researchers have interviewed boat builders, observed boats under construction and in use, and even commissioned and participated in boat building and operation. Archaeologists have also made exact scale replicas of ancient boats, as well as computer reconstructions, based on all of these sources of information (Vosmer 2000).

The main goal for archaeologists in re-creating past objects and techniques is to test the proposed reconstructions and refine construction techniques, as well as to gather new insights about the use of the objects. Experimental reconstructions illustrate gaps in knowledge and design flaws not envisioned until the actual construction and operation of the object is attempted. An essential aspect for an *archaeological* project is the constant checking between experimental reconstructions and the archaeological materials, in a cycle of research, reconstruction and testing. This is one of the central objections to many non-archaeological reconstruction projects, because simply to reconstruct a plausible and workable reed boat is no guarantee that this was the sort of boat made in the past. For example, Heyerdahl (1980) constructed and sailed a reed boat around the Persian Gulf, using construction techniques derived from modern reed boats made both in Western Asia and in South America. However, Cleuziou and Tosi (1994) were able to determine from examination of the archaeological bitumen finds that Heyerdahl's reconstruction, while an effective seagoing craft, was not constructed in the way reed boats were actually built in this region in the past.

As currently reconstructed, the hulls of the reed boats of Mesopotamia, the Persian Gulf, and the Arabian Sea were made by tying together 20 to 30 centimeter thick bundles of rushes or reeds (*Typha* and/or *Phragmites* spp.), inserting a frame, covering the exterior with a reed mat, and waterproofing it with bitumen. Vosmer (2000) describes the probable materials and production processes in detail. The reeds would be cut and dried, then lashed together to form tapered bundles by winding fiber or split reeds around each bundle in a spiraling fashion (Figure 5.2). Bundles were then joined together with thicker twine or rope, perhaps made of palm fibers, to form a boat with characteristically curving ends. From careful study of bitumen impressions, Vosmer suggests that a smaller "interstitial" bundle was placed between two large bundles (Figure 5.2). This would create a stronger and more watertight hull. A frame made of wood or reed bundles was inserted into the hull interior and lashed in place, to maintain the hull shape and provide stiffening. The exterior of the bundle hull was covered with woven reed mats to form a streamlined and easily replaceable outer covering, which was then waterproofed with a coat of bitumen mixture, one to three or even up to five centimeters thick (Vosmer 2000; Cleuziou and Tosi 1994). Bitumen was also used directly on and between the reed bundles, beneath the outer reed mats (Cleuziou and Tosi 1994: 750). Vosmer further describes possible mast, sail, steering and frame arrangements, but notes that there is little direct evidence for these aspects of the reed boats. These reed bundle boats could be quite large, based on ancient textual references to their weights. Vosmer's (2000)

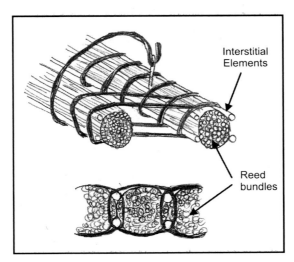

FIGURE 5.2 Close-up of reed bundle construction for Arabian Sea boat. (Redrawn after Vosmer 2000.)

computer and scale model reconstructions were of an average-sized seagoing boat of 13 tonnes displacement, for which he calculated a length of 13 meters and a width (beam) of about 4 meters. It would be able to carry some 5 tonnes of cargo, and easily achieve a speed of 6 knots.

The bitumen mixture was probably derived from liquid seepages rather than hard asphaltum, as it would melt at a lower temperature (Schwartz and Hollander 2001), conserving fuel. Based on analyses of archaeological finds, the bitumen coating was composed of bitumen, tallow, inorganic material, and a large amount of vegetal material. Chemical analysis of some of the Ra's al-Junayz bitumen pieces showed the addition of small amounts of tallow, to aid plasticity, and significant amounts of inorganic material, including calcium carbonate ($CaCO_3$) and gypsum (calcium sulphate), to harden the material and increase its impermeability to water (Cleuziou and Tosi 1994: 754–5, referencing unpublished chemical reports by G. Scala). This paste was then mixed with large amounts of vegetal matter, primarily the same material used to make the boats, *Typha* reeds, but also bits of swamp vegetation, palm leaves, and rarely, straw, barley seeds, date kernals, and *Phragmites* reeds (Cleuziou and Tosi 1994: 754, referencing unpublished reports by L. Costantini). As Cleuziou and Tosi indicate, this vegetal temper would have decreased the weight of the bitumen coating considerably, a very important consideration for a boat. It would also have made the coating easier to apply and maintain, by increasing the plasticity and adherence of the bitumen. When the boat was overhauled or scrapped, the bitumen was removed and recycled, as evidenced by the barnacle bits found in the matrix of some pieces. In addition to salvaged bitumen, which was melted and formed into cakes for storage, fresh bitumen was traded extensively around Western Asia in pottery vessels. Both the bitumen itself and the pottery vessels have been sourced using various analytical methods, and illustrate the extensive trading networks in place in this region at least since the fourth millennium BCE, and probably much earlier (Schwartz and Hollander 2006; Cleuziou and Tosi 1994; Méry 1996, 2000).

The importance of these finds at the small Omani site of Ra's al-Junayz is not only their contribution to our knowledge of the technical aspects of ancient boat building. These insights into reed bundle boat building illustrate some unexpected conceptual similarities between the construction of reed bundle boats and the sewn wooden plank boats also developed in this region. These finds thus raise the issue of the relationship between construction of the reed boats and development of the wooden plank boats. But as Cleuziou and Tosi (1994) stress, the most startling result of their research is the degree to which this small fishing village was incorporated into a long-distance multinational network of trade. The bitumen and perhaps the reeds themselves were exported to the Arabian Peninsula from Mesopotamia; copper fish hooks and all other copper items came from sources in inland Arabia; either the clay or pottery itself for everyday use was made elsewhere; and plant food could be grown no closer than 40 kilometers away (Cleuziou and Tosi 1994). In exchange, fish and perhaps cargo hauling must have been the major sources of local income. Shell products were also traded from Ra's al-Junayz to the Indus Valley and the west coast of India. As Cleuziou and Tosi note, there must have been a great deal of occupational specialization and economic interaction throughout this region, even in village settlements, and a social structure must have been in place that encouraged such interaction and specialization.

It is not surprising, then, that reed bundle boats continued to be used for both riverine and sea transport, long after wooden plank boats came into use. Although the wooden plank boats could be much larger, last longer, and were perhaps more seaworthy, wood was a scarce commodity in this region. The ability to make use of comparatively plentiful materials such as reed bundles allowed more local fishermen to support themselves than would have been possible if only wooden boats were built. And the ability to make relatively large and seaworthy boats of reeds would have added many more boats to the widespread water-based trading system of this region. It would thus have been more difficult for any one group to monopolize trade through the monopoly of transportation, although anyone who controlled wood sources might certainly have dominated trade systems. Archaeologists can use the reconstructions of ancient technologies to understand such economic competition and its effects on past societies. Ancient technologies also sometimes played pivotal roles in the distribution of power within and between social groups, affecting social status and political structure. Another study of reed and wooden boat use illustrates such social aspects of new technologies.

RECONSTRUCTING REED AND PLANK BOATS AND EXCHANGE NETWORKS IN COASTAL SOUTHERN CALIFORNIA

On the other side of the world, based on ethnographic records thousands of years later, the Chumash of coastal and island southern California also made watercraft of reeds and a natural petroleum tar-like substance, but with their own characteristic construction methods (Hudson, *et al.* 1978; Hudson and Blackburn 1982; J. E. Arnold 1995, 2001; Gamble 2002) (Figures 5.3 and 5.4). In contrast to the Near Eastern case, almost all of the evidence for reed-bundle boat use and construction methods comes from ethnographic accounts after European contact, accounts that mention the reed-bundle ("tule balsa") watercraft used by California coastal groups. In addition to this material, ethnographic and experimental reconstructions of reed-bundle boats have been made, the best-known an example of a reed boat made by unknown

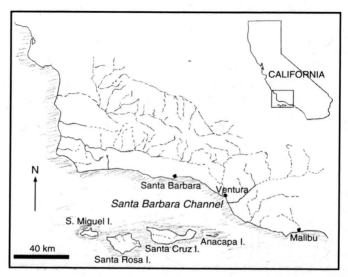

FIGURE 5.3 Map of Southern California Chumash area, showing islands. (Redrawn after Arnold 2001.)

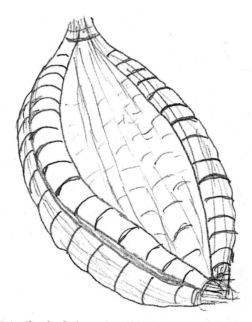

FIGURE 5.4 Sketch of Chumash reed boat. (Redrawn after Arnold 2001.)

Chumash informants for the ethnographer J. P. Harrington in the early 1900s (Hudson, et al. 1978: 20 ftnt 28). Unfortunately, the only documentation that has been found of this reed bundle reconstruction are a series of photographs, in contrast to the extensive notes on the wooden plank boat made for Harrington around 1914, as described below. Later experimental replications begin with the reed bundle boat made and used at the University of California, Santa Barbara in 1979 (Hudson and Blackburn 1982). However, the vast majority of the small amount of archaeological information about Chumash boats relates to the wooden plank boats, rather than the reed boats. Increasing attention to the history of boat construction is changing this situation, as seen in Des Lauriers' (2005) recent article on watercraft used in Baja California.

The primary ethnographic account relating to Chumash reed boats, as told to J. P. Harrington by Fernando Librado, mentions that three types of reed boats were made: "(1) a three-bundle balsa; (2) a five-bundle balsa; and (3) a seagoing balsa in which tule bundles were used like boards" with cracks between the bundles filled with asphaltum mixture (Hudson, et al. 1978: 28). All of these types of reed boats were used in the ocean and estuaries, although they were said to be slow compared to wooden plank boats. Unfortunately, no further information is given about the last type, which may have been much larger. In fact, the editors of the volume give the opinion that Harrington may have read about a unique example of a seagoing multi-bundle reed boat in another ethnographic report, and not actually heard about it from his informants (Hudson, et al. 1978: 31, ftnt 35). The best-described reed boats are the small three-bundle craft. Based on experimental reconstruction and ethnographic evidence from elsewhere in California, they were probably only some 3.3–6 meters (10–18 feet) long, and paddled by only one or two people (Hudson, et al. 1978; Hudson and Blackburn 1982).

A small reed boat could take only three days to create, from cutting the reeds to use of the boat (Hudson, et al. 1978). Reeds, in this case *Scirpus* spp., were cut and dried for several days. The partially dried reeds were formed into three or five large tapered bundles, with a willow pole inserted into each bundle to stiffen it. These bundles were bound up with red milkweed (*Asclepias californica*) fiber string, starting at the center of the bundle and wrapping outward to the ends; sinew was not used, as it was said to rot (Hudson, et al. 1978: 29, 53–54). The side bundles were then tied to the bottom bundle with more milkweed fiber, starting at the ends and working along the sides. As the bundle forming the base of the boat was longer and thicker than the others, the boat curved up at prow and stern. For a five-bundle boat, two bundles would be tied vertically above the lower side bundles, and one or three wooden braces might be tied crosswise, between the side bundles. At least in the Chumash region, the exterior of the boat was then coated with a waterproofing based on asphaltum mined from a few seepage deposits on the Santa Barbara Channel coast, mixed perhaps with pine pitch (Hudson, et al. 1978). The exposed surface was rubbed with powdered clay to cover the sticky asphaltum. No sails were used for these small boats; rather, they were paddled with a double-bladed paddle. If kept out of the water when not in use, the reed boats could be used for a relatively long time, but otherwise they became waterlogged and rotted. In 1979, graduate students at the University of California, Santa Barbara, produced an experimental reconstruction of the common three-bundle reed boat. It was 6 meters (18 feet) long and 1 meter (3 feet) wide (beam), and could carry two people and 64 kg (140 lb) of cargo. The builders were able to collect materials and construct the boat in just 89 hours, and successfully navigated up to 21 kilometers (13 miles) along the coast, but they found the craft slow (Hudson and Blackburn 1982).

These widely used reed-bundle watercraft were overshadowed, in the European ethnographic accounts as well as in Chumash society, by the faster, larger, and more durable sewn plank boats (*tomol* or "plank canoe") that were made only by the Chumash and their neighbors (Figure 5.5). Most of the information about the sewn plank boats also comes from ethnographic accounts after European contact, and particularly from a fortunate collaboration of J. P. Harrington and Fernando Librado on the ethnography and experimental reconstruction of a *tomol* in 1914 (Hudson, et al. 1978; Hudson and Blackburn 1982). However, there is increasing use of archaeological data to document the production history of the sewn plank boat and its social importance. Archaeological studies of plank fragments, bitumen pieces, and chert drills, as well as boat replicas in burials, and data about the development of the Chumash maritime economy and status hierarchies, have provided additional information about the antiquity and construction of the sewn plank boats. Experimentation with sewn plank boat construction dates back to at least the mid-first millennium AD (ca. 400–700 AD), possibly earlier (J. E. Arnold 1995, 2001; Gamble 2002; Hudson, et al. 1978: 22–23, ftnt 6, quoting C. King). Hudson et al. (1978) suggested that the Chumash *tomol* was developed from dugout canoes, based on the form of the plank canoe. As dugouts were not very stable except in estuaries, use of the sewn plank boats would likely

FIGURE 5.5 Sketch of Chumash *tomol*, a sewn wooden plank boat or "plank canoe." (Redrawn after Arnold 2001.)

have rapidly replaced dugout craft in channel or coastal waters. While reed boats appear to have been more stable, and were used in coastal waters, they were definitely slower, clumsier, and could carry much less than a wooden plank boat. (But see Des Lauriers (2005) for drawbacks of the *tomal*.)

There were several types of sewn plank boats, which ranged in length from between 3–6 meters (10–18 feet) up to 8–10 meters (25–30 feet) (Hudson and Blackburn 1982: 345). These craft were made by piecing short planks of wood with V-shaped holes, then sewing the planks together using milkweed fiber string. Split reeds were used to caulk the seams, which were then coated with asphaltum mixed with pine pitch and other substances as waterproofing. The finished boat would be coated with a red ochre and pitch mixture, to seal the wood so it would not absorb water. It might also be decorated with shell inlay or abalone shell "spangles." (See Hudson, *et al.* (1978) and Hudson and Blackburn (1982) for more details.) The sewn plank craft were made of 'patchwork' wood planks because wood suitable for boat-making was scarce in this region, particularly on the Channel Islands themselves. In fact, the preferred wood, California redwood, could be procured only as driftwood logs, making it a very valuable commodity. While the driftwood was scarce, a completed wooden plank boat could last years. In contrast, the reed-bundle boats became waterlogged if left in the water for more than four days, perhaps requiring frequent replacement of these reed boats. Under these circumstances, reeds might actually have been a more limited resource than driftwood, especially on the islands where marshy areas were rare (J. E. Arnold 1995). Building sewn plank boats, although requiring a major expenditure of time and materials at the beginning, may have been more efficient in the long run.

Nevertheless, it is significant that the Chumash wooden plank boats do not replace reed-bundle boats, and there were still reed boats in use during the period of European contact (Hudson, *et al.* 1978; Hudson and Blackburn 1982). The reed-bundle boat was the most commonly used boat along the California coast in general, and the scarcity of historic references to it in the Chumash region may be due to the greater attention given to the sewn wooden plank boat (*tomal*), which was made only in this region (Hudson, *et al.* 1978: 27, ftnt 23). This is a common pattern for new inventions, whether new types of boats or new styles of pottery. Old versions often continue to be used to some extent, serving different functions and requiring different investments of time, skill, and labor. Manufacture of a sewn plank boat did require a limited resource, wood, and more importantly, required specialized knowledge and a large investment of time. Not everyone could afford such a boat, including many people who needed small near-shore craft for fishing or transport. As in ancient Arabia, reed-bundle boats would be an important resource for these individuals.

The development of the Chumash sewn plank boats is also a good case study for the complex way in which environmental conditions can influence technological development. On the one hand, the presence of the offshore islands of the Santa Barbara Channel provided protected waters and an impetus for water travel, both for fishing and transport, so that new types of watercraft could make a significant difference in acquiring resources (J. E. Arnold 1995, 2001). However, environmental conditions do not necessarily determine choices, as is seen in the inverse relationship between the availability of wood for boats and the areas where wooden boats were made in Southern California. The sewn plank boats, preferably made of redwood, were only made south of the area where redwood forests are found, and not in the areas where suitable wood was plentiful. In this case, the desire to travel between the coastal islands and the mainland was a more important impetus to the development of more seaworthy craft than the availability of superior materials.

Manufacturing techniques and construction styles are only a portion of technological studies, though. Technology also includes information about the specialized knowledge and organization of the people making these boats. For example, are there specialized boat makers, or can most people in the society make a boat? Does one person make a boat from start to finish, or are a number of people needed, each of whom has a particular task, from cutting the reeds or making the wooden planks, to preparing the waterproofing mixture and coating the boat? For the Chumash case, we have some ethnographic evidence, at least for the sewn wooden plank boats, that certain individuals were particularly known as specialists in preparation and application of the various asphaltum mixtures (Hudson, et al. 1978). Moreover, plank boat makers were part of a special craft guild, the Brotherhood of the *Tomol* ("wooden plank canoe"), with social and economic responsibilities shared with the other guild members. From the ethnographic accounts, we also know that the entire process of *tomol* manufacture, which took several months to make, was overseen by a master builder. Whether or not there were also master builders of reed-bundle boats for the Chumash is something we do not know. Small reed-bundle boats were widely available, and took only a few days to make (Hudson, et al. 1978), so many people probably knew how to make these relatively simple craft. However, if larger seagoing reed-bundle boats did exist, they may have required specialized building knowledge.

Furthermore, technology studies involve an understanding of the role of these objects in a society, including the status of the craftspeople and the importance of the product to the society as a whole. This sort of information is a key element in determining how highly an object was valued in a society, and in understanding why new inventions are accepted by societies, and why older objects might continue to be used. For example, asphaltum preparation and application does not seem to have been any more or less prestigious among the Chumash than any other aspect of boat-making; there is no additional (or lesser) status involved with being an asphaltum expert. However, the overall director of the *tomal* production process, the master builder, did hold considerable prestige for his knowledge, both within his craft and in society in general. Social prestige was also due to the considerable wealth master builders might accumulate, as they were probably also able to subsequently own and operate the boat for fishing and especially for conducting trade between the mainland and islands. This last was the real advantage of the wooden plank boats over the reed-bundle boats, as the wooden plank boats were swifter and more reliable in the open water, and capable of carrying a much larger cargo, up to 2 tonnes. The commissioning of a wooden plank boat, by those who held enough wealth or labor resources to do so, was therefore a way to increase their wealth and prestige even farther.

For the California case, we know much about the trade networks from ethnographic accounts, and also from archaeological finds of trade goods from known locations. For example, island shell beads are found in mainland residences and graves, and plant food items from the mainland are found on the islands. Similar proveniencing studies of trade items are used to reconstruct trade networks in Western Asia. As noted above, analysis of the asphaltum or bitumen itself provides an additional level of information about exchange networks. Schwartz and Hollander (2006) were able to determine that the bitumen used at the site of Hacinebi in northern Anatolia (modern-day Turkey) came from a number of sources in Anatolia, as well as a source in southern Mesopotamia (modern Iraq). This bitumen was used for sealing and waterproofing a variety of containers as well as boats, so that it was an important commodity, and clearly widely distributed. In contrast, Des Lauriers (2005) suggests that the tar (asphaltum) used by the Chumash was a more scarce resource than the driftwood planks; in this case, asphaltum was apparently not traded.

Arnold (1995; 2001) has suggested that the employment of the larger, faster, more stable wooden plank boats played a key role in the development of unequal wealth and status among the Chumash. Control of the production of these boats, whether through knowledge of how to make them or the necessary wealth to provide the planks, asphaltum, and labor, was one source of wealth, sociopolitical status and power. With the larger size and greater stability of the wooden plank boats, boat owners could accrue prestige and power by increasing their wealth through cross-channel trade, by providing a means of transporting more people to larger gatherings, and by collecting more information through frequent travel by themselves or their agents. Thus, ownership of wooden plank boats was an important mechanism by which social hierarchies could be enhanced. By making transportation and communication networks more reliable, the wooden plank boat likely played an important role in the ability of leaders to extend their power across larger regions. However, this is not an inevitable path of events—the development of wooden boats in other regions of the world does not seem to have played such a role. Restricted control of this transportation source would have been a major factor in the development of Chumash social complexity (J. E. Arnold 1995). As Arnold stresses, the key element in development of Chumash hierarchy was not the development of advanced boat technology, but possible restriction of boat ownership to relatively few individuals. She cites coastal Papua New Guinea as a counter example, where many people owned and operated sea-going boats. New technological inventions themselves do not create social change. Rather, it is the ways in which the members of the society use new technologies that can result in social change, as is discussed in the following section on invention, adoption, and the organization of labor.

INNOVATION AND THE ORGANIZATION OF LABOR

> In a paper on the history of agriculture, J. T. Schlebeker has identified four elements that make for technological innovation: *i* accumulated knowledge; *ii* evident need; *iii* economic possibility; *iv* cultural and social acceptability.... Nor should Schlebeker's list be regarded as exhaustive....
>
> (White 1984: 21)

As Torrence and van der Leeuw (1989) explain so cogently, it is not surprising that there has been much less archaeological work than one might expect on the topic of technological innovation *per se*, between fears of theoretical baggage from our cultural evolutionary past, and the tendency to focus on innovation as an event related to progress rather than as a long-term process involving both change and continuing traditions. There are also often practical difficulties for archaeology in seeing the relatively rapid process of invention, particularly for inventions that are not widely adopted. Neither is it tremendously surprising that there has been relatively little further discussion of innovation as a process in archaeology since 1989, given Torrence and van der Leeuw's astute and comprehensive portrayal of the great complexity of the topic. Nevertheless, the discussion of innovation is necessarily an essential part of a discussion of technology, and van der Leeuw and Torrence's (1989) edited volume is particularly useful in presenting both cases where inventions are adopted, and cases where inventions are rejected and existing traditions of process remain (see also Moorey (1994:v-vi) for Mesopotamia). In this section, I will also explore both adopted and rejected technological inventions, as they relate to the organization of labor. Following Torrence and van der Leeuw (1989: 3), the term *invention* will be used for "the original conception of a new idea, behaviour, or thing" (a relatively rare process), while *adoption* refers to "the behaviour and actions involved in both the acceptance of and use of what was invented." Adoption can refer to both acceptance and use within the society where an invention is developed, and acceptance and use of an invention from outside of the society; the latter is sometimes called "diffusion." *Innovation* is then the entire process from the conception of an idea (sometimes separated off as "discovery") and its realization, to adoption and incorporation within a society with whatever social, economic, or political changes are required (Torrence and van der Leeuw 1989).

THE CASE OF THE GRAIN HARVESTING MACHINE

By looking beyond invention to adoption, it is possible to learn some rather surprising things about the process of innovation. I indicated in the Preface that I have avoided technological examples from Greek, Roman, Medieval European and other cases with substantial textual data, such as China and Egypt, as these cases require rather different approaches than a largely archaeologically focused case. Furthermore, discussions of inventions and technological systems from these societies and time periods have often been relatively well-published for general and scholarly audiences (e.g., White 1984; Cotterell and Kamminga 1990; Gies 1994; James and Thorpe 1994; numerous articles in the history of technology journal *Technology and Culture*). For this example, however, I will have to "cheat" and turn almost entirely to historical data to illustrate a case where innovations that were clearly improvements in terms of labor efficiency were *not* widely adopted, until the social, environmental, and economic settings were appropriate. One of the great topics addressed by historians of technology and economics has been the development of the mechanized grain harvester in the 1800s AD/CE. Most people would be surprised to learn that a working grain harvesting machine had been invented and used nearly eighteen hundred years earlier in Roman Gaul. True, the Roman example was powered by a donkey or ox, not a combustion engine, but the concept of increased labor efficiency through the use of multiple shearing blades rather than individually wielded scythes or harvesting knives was essentially the same. What happened? Why was this hugely labor-saving agricultural innovation never widely adopted? Why did it completely disappear rather quickly, and individual human harvesters remain the norm? In fact, the issue of saving labor seems to have been one of the crucial issues for both the rejection and the eventual widespread adoption of such a device, but not the only issue; social, environmental, and economic factors all played a role.

James and Thorpe (1994: 387–389), drawing on White (1984), Thompson (1952), and various Roman authors, discuss the grain harvester developed and used on the large Roman estates in Gaul in the first century AD. They cite a description by Pliny in AD 77, a description by Palladius in the fifth century AD (Thompson 1952), and most importantly, actual depictions of the harvester in relief sculptures on local Gallic tombstones. White (1984) himself is the best source of information about this invention, as an agricultural specialist for the Classical period, and he describes two versions of the harvester, a smaller single-axle cart made of a frame holding a suspended bag and pushed from behind by a donkey, and a larger single-axle cart with a solid deep body, pushed by an ox. In both cases, rows of knife blades or "teeth" were set on the front edge of the cart at a level to cut just below the heads of grain, and the grain fell into the bag or body of the cart as the animal pushed the harvester, guided by one or two human workers. White (1984: 30, 60–62), James and Thorpe (1994: 388), and Hodges (1970: 200) all provide drawings of possible reconstructions of this grain harvester, and White and Hodges also provide photos of stone reliefs showing these harvesters in use. (White (1984: 53, 174–175) notes that Hodges' reconstruction drawing showing the use of two oxen is incorrect, as are many other published reconstructions.) All of these authors observe that the harvester was not widely adopted in the Roman Empire, despite the fact that running an estate was a major occupation and interest for the Roman upper classes, and that there were numerous manuals on efficient farm management, indicating that efficiency was indeed a concern. However, quite different explanations are provided for this lukewarm reception.

In his regionally focused volume, *Technology in the Ancient World*, Hodges (1970) discusses the Roman grain harvester or "reaping machine" as one of a number of Roman technologies. He states that in general, Roman agricultural systems were focused on self-sufficient estates, with no reason to increase their production for outside exchange. He suggests that the development of the Gallic reaping machine represents an exception, a case where "perhaps" there was both a labor shortage and a short period for taking in the harvest, to encourage a way of increasing production. He notes that the machine was probably inefficient, and probably could only be used in large, level fields, but that it was "symptomatic of the whole of the Roman period that it was not developed to become more efficient and did not achieve more widespread use" (Hodges 1970: 199). In other words, there is a vague notion that the general mindset of the Romans as a group was somehow resistant to innovation, a thread that runs through his discussion of the Romans but is never clearly explained. Hodges, the giant of Chapters 3 and 4 for his masterly summary of so many different production processes in *Artifacts*, unfortunately is much less equipped with the necessary background to assess technological systems rather than production processes (as differentiated in Chapter 1). White (1984: 7) makes a similar criticism of Hodges' *Technology in the Ancient World*, noting his lack of background in assessing the textual material, but at the same time critiquing fellow historians' inattention to technical information. A more general problem is Hodges' vagueness about social processes, which is very much a product of his times for archaeological investigations of technology. The fact that I *expect* a clear explanation of why and how a society might be resistant to the adoption of apparently beneficial new techniques, as is done in several of the articles in van der Leeuw and Torrence less than two decades later, is an encouraging reminder of how far archaeology has come in a very short time.

White (1984) discusses the development of the grain harvester in the context of other Roman agricultural innovations, such as a balanced sickle, scythes, new plow types, and a new type of threshing machine. He also provides considerable detail about the technical problems that had to be overcome for successful operation, and how these were addressed (White 1984: 60–62). White emphasizes the need to consider economic and social conditions in assessing innovations, as indicated by the quote at the beginning of this section, and also indicates the importance of environmental conditions. For the grain harvester, he quotes Palladius to emphasize the great efficiency and economic advantage of these machines for Gaul, but also specifically points to its appearance in northeastern France in an area with large, flat, open fields, in parallel with the first mechanical harvesters of the 1800s (White 1984: 62, 10). Furthermore, White (1984: 10) highlights the importance of considering traditional planting methods, indicating that a large-field harvester would not be of use in Italy where sown grain crops were typically inter-cultivated with planted crops of different types. This is a very significant point. The desire or need for a harvester would thus have had to be strong enough to completely rearrange the entire cultivation system, and this could have unforeseen effects of all sorts, from increased risk of crop failure with a heavier dependence on grain crops, to increased pests in mono-cropped fields, to potential needs for changes in land ownership or exchange systems if a farm did not own enough land to have multiple large fields. In general, adoption of new ideas, techniques, or things is easier if they fit rather easily into existing traditions; the more adjustments to an existing technological system that have to take place, the more passive or active resistance there is likely to be to adoption.

James and Thorpe (1994) concentrate heavily on labor issues in their assessment, suggesting that the use of a slave economy is why the harvester was neglected. They allude briefly to the economic demands, environmental conditions, and labor shortage issues for Roman Gaul that Hodges also raised to explain the initial adoption of the grain harvester: the need "to cope with the large demand for grain in an area with unpredictable weather during the short harvest season and a local shortage of agricultural labor" (James and Thorpe 1994: 387–388). But their chief concern is an explanation for the lack of adoption of this innovation elsewhere (and presumably the eventual abandonment of the harvester in Gaul as well). James and Thorpe propose that the heavy use of slave labor on agricultural estates created social conditions that prevented the adoption of a more efficient harvesting method if it meant that slaves might be displaced by harvesting machines, due to fear of social upheaval. Agricultural slavery was life under comparatively good conditions as slavery goes, certainly compared to the life of a slave in a mine, as noted in Chapter 4, or even life as a factory worker in the city slums of later industrial Europe. Displacing workers from the fields thus raises the possibility of discontented slave riots and revolts if less labor were needed and slaves were in danger of being sold or even losing all means of support. James and Thorpe (1994: 389) suggest a parallel for China, where a push-scythe with a single large blade was described in 1313 AD/CE, but which they propose was not adopted due to "the attitude taken by most Chinese bureaucrats toward laborsaving devices in agriculture"; in brief, that labor was plentiful, so that there was no need for labor-saving machines, and widespread use of machines would have deprived the peasants of work, again raising the specter of revolts. Although James and Thorpe do not suggest it, comparisons might also be drawn with the riots associated with mechanization and loss of livelihood in early modern England.

James and Thorpe conclude with a description of how John Ridley saw a reconstruction of the Roman grain harvester in 1825, and used the idea to construct a mechanical grain harvester in 1843 when faced with a severe labor shortage in Australia. The history of these mechanical harvesters of the 1800s offers some intriguing insights for the earlier case in Gaul. Basalla (1988: 151–154) provides an overview of a number of historical studies concerned with the eventual success of the McCormick mechanical grain harvester in North America. As in Hodges' critique of the efficiency of the Roman grain harvester, Basalla notes that McCormick's mechanical reaper faced a number of mechanical problems, especially with (1) adjusting to different terrains and (2) being simple enough for an average farmer to use and maintain. However, he cites the economic historian Paul A. David's claim that the real reason the McCormick reaper was slow to be adopted was that for small farms, below ca. 50 acres, it was more efficient for the farmer to use scythes than to buy and maintain a mechanical reaper. In other words, economic feasibility was the primary barrier to adoption. David suggests that the reaper gained ground after the 1850s because labor costs rose (no doubt even more so with the American Civil War), the size of grain farms increased, and the reaper cost remained stable. Basalla elaborates on this, particularly with regard to the increase in large grain farms with the expansion west into the prairies, very flat, featureless, rock-free land ideal for a mechanical grain harvester. He also notes the rise in the railroads would encourage surplus grain production, as it could easily be transported in bulk. Although Basalla does not mention it, lower population densities in the prairie regions would make labor even more scarce. Thus, environmental, labor supply, and economic conditions all provided an impetus to the adoption of this invention; conditions were favorable for innovation to occur.

While Basalla does not mention the Roman reaper at all, there are interesting congruences in the explanations for the lack of adoption of the harvester in the Roman period and the eventual adoption of a harvesting machine in the 1800s. The environmental conditions favoring both machines were the same, as White indicated; the large, relatively flat fields of Gaul may have made even a crude grain harvesting machine useful, especially if there were labor shortages. Elsewhere, smaller and less flat farms would not have benefited from such a machine regardless of labor shortages, again in both cases. In the Roman case, recent improvements in hand harvesting tools like sickles and scythes might have been preferable to the machine on all but the largest farms, and the grain harvesting machine would be useless where intercultivation of crops was typically practiced. Labor supply was a major concern in both cases, where labor shortage combined with large farms were key to the adoption of a machine. Finally, economic conditions of both demand and transportation are thought to have affected the North American case, with the building of the railroads providing an immense market in the populated cities of the East for farmers on the plains far to the west. Rome's traditional breadbasket was Egypt, and large-scale grain transport could be accomplished by boat up the Nile and across the Mediterranean during favorable weather seasons. The conditions in Roman Egypt mitigating against the adoption of a grain harvesting machine are a point which specialists in the region might find interesting. The lack of easy large-scale transportation methods for grain from Gaul to large population centers in need of grain may be one piece of the puzzle, but the continued shortage of labor in the North American case may have been a more important contributing factor to the different historical trajectories. Even in these two cases, with their many similarities, we see that explanations for particular situations of innovation require examination of a large picture, where labor supply, environment, social conditions and traditional practices all affected decisions about the adoption of innovations.

DIVISIONS OF LABOR, WOMEN'S ROLES, SPECIALIZATION, AND MASS PRODUCTION OF POTTERY

Innovation is not only about the invention and adoption (or not) of new things, however, as many of the authors discussed above make clear. Innovation can equally relate to the reorganization of the labor force itself, in terms of the people in that labor force, their social status, or the place where work is performed. Much of the archaeological research into these sorts of innovations is discussed in the massive archaeological literature on craft specialization. As a start, Costin (1991) is a seminal summary and analysis of archaeological approaches to craft specialization for all crafts, and Chapter 2 of Sinopoli (2003) summarizes much of the subsequent debate and discussion. One aspect of craft specialization and the division of labor that has generated considerable attention recently concerns the changing role of women in production in various societies. The incorporation of women into the specialized nondomestic workforce has seen dramatic shifts in Europe and North America during the past century or two. There were clear social class divisions associated with women's roles in the late 1800s and early 1900s, when women and children of the lower classes were essential components of the specialized workforce, from domestic labor to factories to mines, while upper and middle class women were often restricted to teaching girls, regardless of their economic situation. World War II saw a very dramatic shift in the role of women workers in many of these countries, with massive enrollment of women of all classes in traditionally male jobs including heavy industry, and this situation was as dramatically reversed after the end of the war. The subsequent slow but increasing enrollment of the majority of women in the nondomestic workforce up until the present day has had major political and social as well as economic effects on the roles of all genders. Similar shifts in the acceptable roles of women in craft production have taken place in the past, in times of war and peace, in textiles, metal working, and trading, both as independent workers and as co-workers with their husbands or sometimes taking over the trade as widows (Devonshire and Wood 1996). In archaeology, the shifting economic and social roles of women have been most frequently examined for textile production, as mentioned in Chapter 3, and for pottery production, especially in relation to the specialization of craft production and sometimes in association with new inventions such as the fast wheel.

The role of women in the "nondomestic" workforce is rather difficult for archaeologists to assess, not only because of the difficulties of obtaining data but also because of the lack of separation of categories such as "work" and "leisure" as well as "domestic" versus "factory" work in many societies in the present, and apparently even more in the past. Pottery workshops are and were frequently part of or immediately adjacent to a potter's house, and the whole family might be involved, even with full-time mass production (van der Leeuw 1977; also summarized in Sinopoli 1991: 98-100). Professional textile workers typically did their work at home in almost all societies until the rise of modern textile mills, although there are at least two exceptions from ancient state-level societies, both with historical records providing information about the changed organization of the craft. Both Wright (1996b; 1998) for southern Mesopotamia and Costin (1996; 1998a) for the Inka of South America discuss and analyze historical accounts showing the varied roles and status of the range of weavers found in both societies. In the Inka case, specialist women weavers had a special status unlike any other craft specialist except female brewers. They were attached specialists removed to state workshops where they lived, permanently removed from their own families and communities. They could not marry, but had a special enforced role as a state virgin, and had high social status but little or no social power. Male specialist weavers and part-time female weavers, some of whom were as skilled as the specialists, did not have these constraints but lived with their families in their own communities, and part-time women even worked at home. In contrast, Wright describes a situation in the third millennium BCE in southern Mesopotamian palace and temple workshops, where the attached weavers, all women and many war captives or slaves, have the lowest pay and presumably the lowest social status of any artisan group even though their products were highly valued. These specialist women weavers did have children who lived with them although they do not appear to have been married; the female children were trained by their mothers to become weavers while the male children were sent out to do menial labor. The male specialist weavers Sinopoli (1998; 2003) describes for medieval south India are more typical of the general pattern, living and working from their family homes, and in fact remarkably free of state control other than taxation. Many of the other articles in *Craft and Social Identity* (Costin and Wright 1998) also refer to the location of work and status of female workers, especially textile workers; Wattenmaker (1998), for example, provides northern Mesopotamian cases with only archaeological data that indicate the presence of textile producers working in homes of low economic status, as well as textile producers working and living in workshops associated with higher economic status. The location of production thus does not equate simply with either status of producers or type of production control; each case requires careful examination of multiple lines of evidence. This should not be seen as a situation encouraging us to despair about our abilities to understand the past. Rather, the results coming from such studies are providing new pictures of the rich complexity of ancient societies that are far more interesting than previously imagined.

The organization of production for fiber crafts also provides some excellent examples of the complexity of characterizing "specialists." Wendrich's (1999) ethnoarchaeological studies of basket production in Egypt, particularly her innovative use of video recording and analysis, led to intriguing finds about the skills held by occasional basket makers producing for their own use, versus professional basket-makers producing for exchange. The skill of the basket makers was not linked to their professional or nonprofessional status; both could have the "skill" characteristics of steady and economical movements in production. However, the speed of the professional producers' work was always fast, and this was not true of the nonprofessional basket makers. Wendrich (1999: 391–393) suggests that the archaeologically-detectable mark of a professional basket maker would be signs of haste in the form of small inaccuracies, but with a skill-based regularity to these small inaccuracies. This detailed research adds significant dimensions to the archaeological debate about what characterizes "full-time" versus "part-time" producers, and whether such a distinction is even important for understanding technological systems. Bril, Roux, and Dietrich (Bril, et al. 2000; Roux, et al. 1995) also used innovative ethnoarchaeological and experimental methods to examine variations in skill, but for craftsmen working with hard stone. They carried out task experiments derived from psychology with bead-workers who had different levels of experience in knapping, and analyzed the resulting products using a pattern-recognition computer program to study the variations. Their work indicated that ten years of practice were needed to become "expert" knappers, capable of producing *all* types of beads. In this case, the type of bead produced as well as the quality of production was a useful indicator of skill level.

Archaeological studies of craft specialization and the organization of pottery production have investigated what types of objects were made, how, and by whom, as well as where and when pottery was produced and who controlled the production and distribution. (See Sinopoli 1991, 2003; Costin 1991, 2001; and Rice 1996b for overviews and further references.) Innovations in the division of labor in pottery production, including gender divisions, have occurred in many societies. Such innovations are sometimes linked to the appearance of newly invented tools or techniques. This does occur, but far more often innovations in pottery production and organization come from adapting or inventing tools and techniques *in response to* new social and economic conditions rather than the reverse, particularly where pottery production techniques are already well-established. A notable example is the connection often vaguely made between the invention of the potter's wheel, "full-time" specialization, and a switch from female to male potters. This particular linkage is based on assumptions and simplifications that are problematical, and the key issue is actually not the invention of the potter's wheel, but the shift to mass production.

First, as is discussed in Chapter 4, and illustrated in Figure 4.5, the term "wheel-throwing" is often used to refer to the use of almost *any* type of turning device in pottery making. Technically, throwing should only be used to refer to true, pivoted wheels rotating at a rapid speed for a considerable period of time (Rice 1987: 132–134), but this is not always easy to determine archaeologically, especially from the examination of a relatively small sample of pottery. Tournettes can be turned rather quickly, producing marks on the pottery characteristic of fast wheels (see Chapter 4). Second, use of the fast wheel is almost automatically associated with rapid production of large numbers of vessels, because of the association that has been noted in numerous ethnographic contexts between such mass production and use of the fast wheel, especially use of the fast wheel for off-the-hump production of vessels. While the fast wheel is frequently used for rapid mass production, it is also used for other purposes. Potter's wheels, including classic "fast" kick-wheels, are frequently used as slow wheels, turned gradually in the building up of large vessels with coils and smoothing, or to scrape or finish pots made by wheel throwing or other methods. The care that needs to be taken to avoid equating any evidence for wheel-turning with mass-production is shown by the recent careful work by Roux and Courty, summarized in Roux (2003), that has provided data on the long use of the wheel in the Levant as a slow wheel or tournette for scraping and careful alteration, prior to its use as a fast wheel. (While the long use of a slow wheel in the Levant was suggested long ago by Johnston (1977: 206) based on ethnographic experience, Roux and Courty provide archaeological evidence.) In addition, there are a variety of methods used to achieve mass production, including molding, rapid coiling on a slow wheel, and a combination of methods, so use of wheel-throwing as a proxy for production scale is extremely problematical, except in very specific cases where the range of vessels produced in an assemblage are fairly well understood. Many parts of the world, including the Americas, clearly created large amounts of pottery and clearly had highly skilled specialist potters, yet did not employ the fast wheel at all. Third, while mass production of pottery is typically carried out by full-time specialists, this is not inevitably the case, particularly where pottery making is a seasonally restricted activity. Even if we avoid the problem of "full-time" specialization, and refer to *craft specialization* as a significant time devotion to the craft by a person with a relatively high skill and experience level, many specialist potters do not focus on mass production of many vessels by very rapid techniques, but create a smaller (but still substantial) number of vessels that require greater skill to produce. It is the documented use of the wheel specifically for rapid mass production of a significant percentage of the entire pottery assemblage that is the key to shifts in production, not the mere invention of a potter's wheel, which can be used for a variety of purposes.

In terms of innovation, adoption of the potter's wheel as a tool for rapid production of large numbers of vessels may require changes to the entire pottery production system on multiple levels. Throwing on the fast wheel places constraints on the clays and tempers which can be used, as large particles in wheel-thrown clay will mar or even tear the pot (Sinopoli 1991: 101), as well as constraining the shapes which can be easily and quickly made. In some places the available clays might be ill-suited for fast wheel production, or might require more investment in time or innovations in processing to remove large particles. Changes to firing techniques might be required to allow the firing of clays without large temper particles. The fast wheel as a tool for rapid production would be less likely to be adopted in such cases than in regions where a fast wheel required few changes, as in the regions of the world with abundant deposits of fine alluvial clay requiring little temper which were already in use for pottery production, such as Mesopotamia, the Indus Valley, and China. It is not surprising that these are all areas where throwing on a fast wheel is adopted relatively early *when economic and social conditions were in place* that encouraged specialists to mass-produce vessels.

The final linkage often made, that of the association of the potter's wheel with male as opposed to female potters, is based on ethnographic and historic data, where indeed there are no significant examples of societies with primarily female potters using the fast wheel. However, I would argue that the association between male potters and the fast wheel is less related to the technical use of the fast wheel itself than to the social and economic conditions encouraging its use, as male potters also typically dominate the craft in societies where large-scale, skilled, and/or professional production of vessels and other objects is accomplished without the use of the fast wheel, as in the Classic and post-conquest Maya (Reents-Budet 1998: 73; J. E. Clark and Houston 1998). Similar associations of male artisans and large-scale professional production of objects is seen for a wide variety of crafts, even in textile production which is (cross-culturally) the most strongly female-associated craft besides food production. The innovation in organization of production that is associated with the adoption of the fast wheel as a method of mass production is more precisely focused on increased specialization in the production of each stage of the production process, not simple male versus female potters. With increased production scales, pottery production no longer involves one potter carrying out all the stages of pottery making, from collection of clays to firing of vessels (see Figure 4.3). Instead, with the need for greater production, there is usually a division of labor between the stages, with the most skilled or experienced potters using the wheel(s), apprentices processing the clay and temper (which now requires more time and care), and sometimes yet other specialists painting the completed vessels. This can take place in a factory-workshop setting, as in porcelain production in China and Europe. However, it can also take place in a family-based workshop, as is frequently the case in modern South Asia, where the specialists using the wheel are the older males of the family, but the specialist painters or the producers of mold-made sections of the pot are often the female members of the family (e.g., David and Kramer 2001; Rice 1987; Sinopoli 1991). The clay might be processed by women, men, and/or children, depending on time and availability. The question of why males so frequently dominate key stages or types of production in these sorts of production situations is one that archaeologists, anthropologists, and others are still striving to answer. This discussion has been about clarifying the *question*, usually a helpful first step.

The central point that reoccurs in studies of innovation and technology is that much of the innovation in technology is not related to increasing efficiency or quality, although this happens also, usually in the early stages of a craft's or technique's development. Rather, much technological innovation is associated with changing social and economic demands and circumstances. This point is discussed further in Chapters 6 and 7. In the next section I discuss another aspect of innovation that has only been mentioned in this section; the process of tradition and the role of technological choice in technological continuity and change. As van der Leeuw (1993) has remarked for studies of pottery production, it is important to investigate both the choices made and the choices not made together, to look for alternatives in the production process and the technological system as a whole and try to explain why particular routes have been taken in technological pathways.

TECHNOLOGICAL STYLE

> In ancient Mesopotamia, traditional ways persisted side-by-side with newer ways...
>
> (Moorey 1994:vi)

STYLE AND TECHNOLOGICAL STYLE

There has been a long tradition of the study of *style* in archaeology, as is masterfully summarized by Hegmon (1992; 1998). Hegmon (1992: 518) pithily encapsulates the major approaches to style in archaeology, commenting that "for Sachett, style bears particularly on time-space systematics, for Wiessner it has communicative function, and for Hodder it relates to cognitive processes." Hegmon's 1992 article should be read for more details and her extensive bibliography on archaeological approaches to style, updated and extended to include the concept of technological style in her 1998 publication. A major watershed in archaeological approaches to style came from the introduction by Wobst (1977) of the idea that style not only reflected information, but also communicated it; in other words, style was an active phenomenon. Recent approaches to style have also shown that there are a variety of types of style in objects and behaviors. Sachett and others writing about isochrestic variations in style are interested in style as choices made between functionally or technologically equivalent alternatives that are characteristic of particular times and places. This approach has been heavily applied to technological studies, as might be expected, and is a major part of most formulations of technological style, as described below. These isochrestic styles can be socially transmitted through formal or informal learning processes, and so can reflect historical traditions and social relations. This time-space conception of style pre-dates communication approaches, but has been updated to include them, including what Sackett refers as iconological style, style focused on the expression of social information (Sackett 1986; Sackett 1990; see Hegmon 1992, 1998 for earlier references). Weissner focuses heavily on the communication functions of style, and her exchanges with Sackett clarified both approaches and led to the recognition of the existence of multiple types of styles in any society at any particular place and time; indeed, in any one object (Wiessner 1990; see Hegmon 1992, 1998 for earlier references). Hegmon (1992) sorts out the many different types of styles that have been identified, and appeals for the refinement of existing typologies rather than the creation of yet more terms. Surprisingly, this has been done to some extent, with continued use and refinement of many of these earlier terminologies, particularly Sackett's and versions of Weissner's terms. Finally, the interest in style as representative of cognitive processes, as ways of doing and thinking, as Hodder and others have put it, has also generated much interest in the most recent studies of style and technology (e.g., articles in Dobres and Hoffman 1999).

Technological style refers to the application of questions about style to the study of technology, and like style it also has been defined in a number of ways. It is best known to archaeologists through the writings of Lechtman, especially her 1977 publication. Earlier formulations of technological style can be found in a joint paper by Lechtman and Steinberg (1979), written in 1973 but only published in 1979, and itself part of a group of ideas about style discussed by scholars at the time (Lechtman 1977: 3–4). It is worth quoting at length from that paper, because it sounds startlingly like "new" ideas published about technology in the 1990s:

> If we claim that technologies are totally integrated systems that manifest cultural choices and values, what is the nature of that manifestation and how can we 'read' it? ... We would argue that technologies also are particular sorts of cultural phenomena that reflect cultural preoccupations and that express them in the very style of the technology itself. Our responsibility is to find means by which the form of that expression can be recognized, then to describe and interpret technological style.
> (Lechtman and Steinberg 1979:139).

Lechtman's 1977 chapter is the introduction to a group of chapters on technology and style in a book on material culture and technology (Lechtman and Merrill 1977), and it is odd that Lechtman's chapter is so frequently referenced in recent archaeological literature, and the others seldom are. In this section I will discuss one of those chapters at some length, Steinberg's comparative discussion of the technological styles of three different metal working traditions in ancient Eurasia. Other chapters in this volume are by Leone on architecture in nineteenth-century American religious utopias, and Adams on a wide range of ethnographically documented technologies in Indonesia; all of these publications refer to various studies by Cyril Stanley Smith and others. An entire volume of the Ceramics and Civilization series published in 1985 is about technology and style, including technological style (Kingery 1985). In short, there was clearly a rich and vibrant community of scholars working in several fields on aspects of what we would now call technological systems. While Lechtman's work is definitely central to the concept of technological style, it is worrying that other work from this time period seems to have been dismissed or forgotten in recent summaries of anthropological approaches to technology, giving the impression that technological systems were discovered in the 1990s, at least for the Anglophone literature. (See Stark (1998) for a very brief review of the contrasting history of developments in the Anglophone and Francophone traditions relating to style and technology.) The rich data and thoughtful conclusions of these earlier researchers are well worth reading and "translating" into current terminology, directly commenting on the weaknesses we have moved beyond (such as the generally passive aspect of style assumed in many of these works), but not ignoring them or we risk losing valuable insights from the hard-won knowledge of our intellectual predecessors. The old expression about reinventing the wheel is something technology specialists might particularly want to avoid.

So what then is technological style? Like style, it has a number of interlocking definitions, including both passive and active non-verbal communication and the manifestation of cognitive processes, primarily as expressed through the choices made in the practice of technological processes where alternative choices exist. In fact, all of these expressions of technological style can be found singly or together in Lechtman's 1977 publication alone (respectively, communication and performance on page 13, attitudes of artisans and cultural communities on page 10, and casting versus forging on page 7). This indicates the complexities of stylistic expression recognized from this early period by technology researchers. Lechtman here is primarily focused on the role of technological style as communication and expression of cognitive processes, which is perhaps why this publication has been so popular. Other works have discussed at greater length the methods of expression of technological style itself, the actual types of choices made which can and have been made (e.g., Steinberg 1977; Wright 1993; Ehrhardt 2005; and many others, see below). In her own work, Lechtman has focused on a technique of production (alternative methods of metal surface coloring) as an expression of technological style in Andean metalwork, more recently elaborating further on other technique choices as types of ethnocategories (Lechtman 1999).

Most studies of technological style have also employed production techniques or the materials chosen as the means to examine *technological style of production*. The majority of the examples in this section will therefore focus on production techniques and materials, in large part accessed through archaeometric study of finished and unfinished objects. While these are in no way the only methods of expression of technological style, they are usually the most accessible. *Technological styles of the organization of production*—the order and location of production stages, or the nature and organization of production personnel, for example—involve knowledge of production traditions that archaeologists do not always have, although there have been a few such studies. In most cases, the authors have tried to place their studies of technological style of production or organization of production within the context of the overall technological system, as defined in Chapter 1.

As noted, technological style is primarily analyzed through the choices made between approximately equivalent alternative options in technological production. The choices can be the types of materials used, the techniques or tools employed, the organization of production stages, the nature and organization of production personnel, and so forth. Characteristic technological styles can relate to functional or economic reasons, as well as social and religious reasons. For example, Hosler's (1994a; 1994b) well-known research into ancient Mexican metals found that the choice of materials used was heavily related to desired characteristics of sound and color inspired by religious and social beliefs, in an analysis that crosses back and forth between investigations of style and technological style. Vandiver and Koehler (1985), in contrast, examined a case where functional and economic conditions were likely the main reasons for changes in technological style. They provide extensive data on the development of three separate technological styles of pottery production for the same type of object, amphora, from seventh to second century BCE Corinth. Vandiver and Koehler (1985) found differences in materials, shapes, and techniques of working. They conclude that the reasons for the differences in the styles are likely to be functional differences in use, some of which may have been communicated by the different shapes, as well as changing access to raw material and/or changing socio-economic demands.

Steinberg (1977) similarly compares types of production methods for three different groups of vessels, examining bronze drinking vessels associated with funerary drinking rituals in three different societies. He notes that the Shang Chinese bronze vessels were cast in very intricate, complex piecemolds, with designs and techniques that are highly *skeuomorphic* (imitating one material with another) of work in clay. The Phrygian Anatolian vessels were also very skillfully made but with a quite different aesthetic, outstanding examples of finely raised vessels whose simple forms deceptively mask the skill required for their extremely uniform, even raising. The Late Bronze Age vessels from

Central Europe were also fabricated rather than cast, but rather than smooth, perfect raising, they are often a bit mis-shapen, raised too quickly and pleated or torn where insufficiently annealed. Their rivets, edges, and seams are all apparent and sometimes a bit sloppy. Steinberg uses these three very different production styles as insights into the working conditions and status of the metal workers who employed them. One of the most interesting aspects of Steinberg's study is his interest in going beyond technology styles of production methods, to look at how the production *methods* might have been influenced by the technological style of production *organization* in each case. He suggests that the Shang Chinese and Phrygian Anatolian metal workers were probably patronized by high-level elites in a highly stratified society, and worked under more permanent conditions than the Central European metalsmiths, who may have been somewhat itinerant and who were probably also producing a broader range of objects (including armor) for a larger group of less-stratified, less exacting elites. Their techniques and the quality of their work would accordingly be rather different. In other words, Steinberg moves beyond categorizing metal working traditions as produced by "highly skilled" or "low-quality" workers to examine what it was about the social or economic conditions that made particular qualities of work acceptable or not. The ancient metal workers are clearly very much alive for Steinberg, and he describes the techniques and products through the eyes of a craftsperson rather than a consumer, but the eyes of a craftsperson who is very aware of the desires of his or her consumer.

TECHNOLOGICAL TRADITIONS: METAL AND BONE WORKING IN NORTH AMERICA

A similar interest in comparing metal working traditions is found in the work of Ehrhardt (2005) and Anselmi (2004), both of whom are interested in the metal working traditions of eastern and mid-continental North America around the time of European contact, both employing archaeometric analysis of metal objects as well as archaeological and ethnohistoric data where available. Ehrhardt (2005) examines the large assemblage of metal artifacts and debris ("scrap") from a Late Protohistoric (1640–1683 AD/CE) Illinois village site in Missouri now called the Iliniwek Village, that has been identified as the historically described site of Peouarea. Anselmi (2004) concentrates on metal objects and scrap from a number of early contact Wendat (Huron) and Haudenosaunee (Five Nations) Iroquoian archaeological collections from northeastern North America in the Early (1480/1500-1614 AD) and Middle (1614-1690 AD) Contact Periods. Both are dealing with times of growing contact with European missionaries and traders in these two regions.

Both provide fascinating information from small bits of corroded metal as they use patterns of production and consumption to understand the technological style of production for these two groups, as they interact with and procure metal objects from Europeans. Similar studies have been carried out by Leader (1991) for the Calusa of Florida. Anselmi's (1994) work examines multiple sites and native groups over more than one time period, and so is able to further show differences in the degrees of adoption of tools and techniques by different native groups over time.

Ehrhardt (2005) draw on the work of Franklin *et al.* (1981), who described an overall North American approach to copper working that was based on fabrication of native copper (working without melting, as defined in Chapter 4), in contrast to metal working traditions in most of the world which employed smelting and melting of metal ores as well as fabrication. The same is true for North American traditions of iron working in the Arctic, as noted in Chapter 4. To use my terminology from Chapter 3, metal working in North America was thus an extractive-reductive, not a transformational technology. Franklin, *et al.* (1981) also discuss regional and temporal variation based on other aspects of production, that is, regional and temporal technological styles. (Following Franklin, *et al.* (1981), Ehrhardt (2005) uses a terminology of primary and secondary 'techniques' that I will not employ, as it is close but not quite the same as the way primary and secondary 'stages of production' are commonly used in most discussions of metal working, as represented by Figure 4.11.) Working as they are in situations of cultural contact, Ehrhardt (2005) and Anselmi (1994) explore the large-scale differences between the European and North American metal working traditions, but also provide significant data about the similarities and differences between the two regional technological traditions, that practiced by the Illinois in the mid-continent Mississippi river drainage and western Great Lakes, and by the Iroquoi in the northeast, to the east of the Great Lakes.

As detailed in Martin (1999), pre-contact metal working in North America was centered on the plentiful supply of native copper available in the Great Lakes region, which was widely traded in the Great Lakes and Mississippi River drainage regions, although Ehrhardt (2005: 59, 69) references studies showing use of native copper deposits from elsewhere in North America as well. Copper was not smelted or melted, but was hammered and annealed into sheets and objects from at least 3000 BCE, with sheet production a major part of the production system after the earliest period. Joining only involved cold-working methods, primarily metal riveting or nonmetal adhesives or binding. Metal pieces were cut using groove-and-snap methods, as was also done in stone working. Some groups made elaborate "cutout" shapes from copper sheet by incising the shape on one side of the sheet, then grinding through the raised projections on the other side, notably the Hopewell people

(200 BCE to 400 AD/CE) and later the Mississippians and participants in the Southeastern Ceremonial Complex (from 900 AD/CE up to European contact in some areas) (Martin 1999; Ehrhardt 2005; Anselmi 2004).

Ehrhardt (2005) describes the changes in technological style for the Illinois with European contact as primarily affecting methods of cutting and joining, while processing methods of forging or shaping by hammering with annealing remain essentially the same, with no interest in melting or casting. However, they no longer create their own sheet from copper lumps, but procure copper and brass already in sheet form from their European contacts. Cutting is now accomplished by shearing with scissors or snips of some kind, or possibly iron knives, based on the characteristic burr marks and folding patterns seen in the metal objects and discards. The cutting method used for a few examples of jagged tears is not clear, but there is almost no evidence for groove-and-snap or even bend-and-snap cutting, nor the more elaborate grinding method described above for cutouts. Although no such tools were found, it appears that the Illinois adopted European shearing tools for cutting, unless it is possible to achieve a shearing burr with a single blade knife; this would be a useful topic for experimental replication and analysis. The other major forming techniques are rolling or bending and perforating; rivets are not used, and only seldom are edges regularly smoothed or traded sheets thinned.

In contrast, Anselmi (2004) found a stronger continuity in the traditional North American methods of cutting in the Iroquoian materials. Although in close contact with Europeans and procuring metal and other objects through European trade, including sheet metal in the form of kettles as well as iron blades, European tools such as scissors or snips were not as commonly used, except by one group, the Mohawk. Instead, there is abundant evidence for use of groove-and-snap cutting followed by grinding of edges to smooth them, techniques typical of the earlier general North American tradition also found in stone working. This difference is particularly interesting as the Iroquoian groups and the Illinois were fashioning much of the European copper and brass into very similar ornamental objects, such as rolled beads, tinkling cones (rolled cone-shaped pendants), and flat pendants. In both cases, the majority of metal objects are ornaments, although there are a good number of both formal and expedient tools found in the Iroquoian assemblages, increasing in abundance through time. Anselmi (2004) also examined metal artifacts made in European settlements at around the same time, and found clear evidence for the use of typical European metal working techniques, with the European assemblages showing significant statistical separation from the Iroquoian assemblages. This continuation of the traditional cutting method in the Northeast is rather surprising as one might posit the likelihood of new, more efficient techniques being adopted along with the new materials. However, Anselmi (1994) does also find increasing adoption through time of European techniques as well, at least for some Iroquoian groups. Anselmi thoughtfully discusses these findings in the context of the long history of anthropological research into cultural contact, in general and for the Iroquoian groups of North American.

As both of these scholars explicitly addressed issues of technological style and issues of cultural contact and change in their detailed analyses of large numbers of artifacts, some sophisticated questions about technological systems can be asked of their research beyond their solid, useful data on technological processes. First, what might be the reasons for the differences in the two regional technological styles, especially cutting methods? After all, at least some of the Northeastern groups did also have access to European tools like snips and scissors. One answer to the differences in cutting methods might be that they had access to different types of raw materials. Ehrhardt (2005: 107) notes that the European sheet metal (both copper and brass) that was used at the Iliniwek Village site was quite different from that found on sites in the East: uniformly flat and smooth without the raising or turning striations found on most European kettle-based sheet, and much thinner. The Iliniwek metal is thus much easier to shear and bend, either allowing or requiring use of the European cutting method–again, experimental studies might be very helpful. This difference in working properties might explain why the one or two tools found at Iliniwek were also among the very few objects made of native copper rather than traded (smelted) copper; the traded metal may simply have not been suitable to form the tool. Alternatively, or additionally, the use of shearing techniques for cutting might have related to a desire for more rapid production at Iliniwek, encouraging the acquisition of a European bi-bladed shearing tool which was not among the normal trade items, unlike single-bladed iron cutting tools (knives). Both groups had significant trading contacts with the Europeans, both merchants and missionaries, and at least some of the Iroquoians did acquire European snips and scissors, so a simple explanation related to access is not sufficient; active procurement of these tools implies a desire for this production style, for whatever reason(s). Hints of a desire for rapid production might be found in the tendency towards less finishing in Iliniwek objects, with many examples of rough edges, shearing burrs, and irregular finishing—what Ehrhardt (2005: 184) refers to as evidence for expediency in the production processes. As Ehrhardt immediately points out (citing a conference paper by Latta and Anselmi), the Iliniwek do not seem incapable of more finished, careful metal work, but rather the expediency of the work does not seem to matter to the consumers. Steinberg's (1977) analysis of the Late Bronze Age Central European metal workers discussed above provides a parallel. It would be interesting to more thoroughly examine the Mohawk case, the Iroquoians who made the most use of European snips or scissors (Anselmi 2004). The abandonment of cutout techniques of cutting by the Iliniwek, in contrast, is almost certainly not about changes in material availability or expediency of production, but in the fact that use of these cutouts are no longer a part of the Ilinewek ideological system, as they are no longer participating in the Southeastern Ceremonial Complex.

Finally, significant differences in the Iroquoian and Ilinewek attitudes toward metal scrap are puzzling to Ehrhardt (2005: 75, 108, 138, 190–191), who expected the same sort of careful conservation and reuse of scrap seen for the Iroquoians, where over 50% of the "scrap" from Wendat (Huron) sites turned out to be metal pieces used as expedient tools based on use wear studies (Latta, *et al.* 1998). Latta, Thibaundeau, and Anselmi (1998: 179) explain this pattern as showing that rather than debitage or wastage, these accumulations of irregular bits of metal served as stores of raw material to be drawn on at need. In contrast, while metal discards from the Ilinewek site often show signs of working prior to disposal and some are even prepared blanks, there is no evidence of curation or reuse. Ehrhardt notes that this could be due to a very regular supply of metal to the Illinois, decreasing the value and status of the metal, but she also notes that this metal could have been traded into the other regions farther from European trading contacts where the Illinois had regular trade contacts. Comparative information from other Illinois sites might help to resolve this issue, indicating how unusual or typical the pattern of discard at Ilinewek Village might be for the Illinois.

To return to the large-scale North American pattern of metal working, I noted that groove-and-snap techniques were commonly used, even after European contact in many areas. An analogous persistence in the use of groove-and-snap techniques in spite of the availability and indeed the use of European tools is found in bone working assemblages from the opposite side of North America. Wake (1999) examined bone artifacts and working debris from excavations at the nineteenth-century fur trading and agricultural outpost of Colony Ross, north of San Francisco on the California coast. Colony Ross was run by the fur-trading Russian-American Company (RAC), and housed Russian colonists, local Kashaya Pomo Native Californians, and Aleut and Qikertarmiut Native Alaskan specialist sea otter hunters hired by the company. The Native Alaskans were recognized as tremendously effective sea otter hunters, which was why the Russians brought them to Colony Ross. The Russians thus had every incentive to provide the sea otter hunters with whatever was needed for their very efficient traditional hunting methods, as is borne out by historical accounts. These Alaskan traditional hunting techniques employed bone arrow and dart points used in compound throwing tools, made by the Native Alaskan community. Wake (1999: 198–199) found that although metal knives, saws, and other tools were made available by the Russians, the bone workers used these tools in traditional North American ways. Specifically, bone shaping was done not by sawing, but by grooving or scoring the bones with a small or medium-sized metal knife around the circumference, then snapping the bone along this line. Even on the rare occasions (three cases) where saw marks have been found on bone working debris, the bone was grooved and snapped, not cut completely through in the European manner of saw use. As Wake indicates, the metal cutting *tools* were completely incorporated into the technological style of bone working as superior versions of existing tools, but the *techniques* and *products* were scarcely changed.

This continued use of bone hunting tools, rather than adopting metal harpoons, is in itself a noteworthy aspect of the assemblage. Wake (1999) mentions that this is the case for most archaeological sites in Alaska as well, with metal harpoons only found after the 1850s, and that there is no mention of metal tool use in hunting any marine mammal in the available historical record. Wake provides a number of explanations that highlight the fallacy of the "metal is always better" attitude toward tool use: bone toolkits were probably easier to maintain and produce than metal tools and bone easier to procure, and bone tools would not corrode in the marine environment. Furthermore, he notes that the use of heavier metal tools might require the acquisition of new throwing patterns for these already master hunters. In addition, Wake mentions that there are ethnographic and historical accounts that metal tools were not spiritually appropriate for hunting marine mammals for Native Alaskans. (Presumably this bar did not extend to the use of metal tools to make the bone hunting tools.) Such stipulations about the use or prohibition of particular materials for hunting particular types of animals is seen in McGhee's (1977) discussion of the use of sea mammal bone and ivory for making tools for hunting sea animals and birds, while antler tools were used to hunt caribou and other land animals, for the historic Inuit and prehistoric Thule of the North American Arctic. Although subsequent research has shown that these associations were not necessarily as strong as found in McGhee's cases, all of these investigations of technological style show that *both* ideological and materialist factors need to be considered.

There are many other examples of differences in technological traditions seen in different technological styles of production. The recognition of different technological traditions of flint-knapping allowed Luedtke (1999a, 1999b) to differentiate historic-period gunflints made by Euro-Americans and Native Americans in northeastern North America, by the different technological styles of flaking patterns. Russell's (2001a; 2001b) analysis of the Eastern European, Anatolian, and Baluchi bone working traditions, described in Chapter 3, is a good example of a case where end-products and working techniques are quite similar, yet the methods of organizing production appear to be very different. Shah (1985) describes three completely different technological systems for making terracotta figurines in Gujarat, India, with three different types of clays, different manufacturing techniques, different types of craftspeople in terms of gender and professional skills, and different organizational systems for production and distribution; yet all of these figurines are used to represent the same objects, spirits, and desires in the same religious ritual system. In the next chapter, I continue this exploration of technological traditions with an examination of the development of new materials in the Indus Valley Tradition, and their possible relation to social communications about status and group identity.

CHAPTER 6

Thematic Studies in Technology (*Continued*)

The discussions of innovation and technological style in Chapter 5 are also related to the topics of this chapter. In the first section, I examine innovations in the materials used to make ornaments and "luxury" goods in the Indus Valley Tradition of South Asia, and how the use and development of these new materials relates to continuities and changes in the technological traditions of this region, particularly in relation to apparent changes in social status and identity. In the second section, I examine the technologies used in religious rituals in the American Southwest, again seeing examples of changing and continuing technological practices. However, additional archaeological approaches and topics are also covered in this chapter, such as the archaeological determination of the relative value of objects for prehistoric societies, and the relationship between valued objects and status. The importance of discard patterns illustrated in the second section shows why technological systems do not end with consumption, and why archaeologists are so precise about the exact structure of heaps of garbage.

VALUE, STATUS, AND SOCIAL RELATIONS: THE ROLE OF NEW ARTIFICIAL MATERIALS IN THE INDUS VALLEY TRADITION

> Speaking of the use of lapis lazuli and turquoise by Harappans, Jean-François Jarrige once said "They didn't like them because they couldn't play with them." In all Harappan craft production, a major emphasis is placed on the creation of artificial substances more than on the employment of precious, well recognizable raw materials.
>
> (Vidale 1989: 180)

In Chapters 3 and 4, I discussed one way to divide material-based approaches to craft production, into "extractive-reductive" and "transformative" crafts. *Transformative crafts* transform raw materials through pyrotechnology or chemical processes to create a new human-created (artificial) material. The vast majority of crafts of this type were *pyrotechnologies*, technologies involving the transformative application of heat as an essential part of production, such as the production of pottery and other baked clay objects, metals, glass, and lime plaster. In this examination of the role of new artificial materials in the Indus Valley Tradition of South Asia, I discuss one aspect of the question "Why were such new materials invented and adopted?"

USES OF ARTIFICIAL MATERIALS

Artificial materials, materials that have had their basic physical or chemical structure transformed by human action, were invented and adopted for a variety of reasons. Many such materials were eventually used for both utilitarian and nonutilitarian objects. The common assumption is that new materials were first employed for tools or other "utilitarian" applications, stemming from the popular belief that technological invention was all about necessity from a food-procuring, shelter-creating point of view. In fact, many of the major classes of artificial materials invented in antiquity were first used for quite different purposes. Here I need to be clear about my use of the term "utilitarian." The narrowest definitions of "utilitarian" refer only to objects related to physical survival in terms of essential food procurement and shelter. Other definitions include additional objects perceived as necessary for physical survival by the people who owned them. For people with biologically-based conceptions of illness, invisible agents such as viruses or bacteria are seen as a primary cause of illness, and so proper sanitation systems and hygienic practices are seen as necessary for physical survival. For people with spiritually-based conceptions of illness, which can be either alternatives or additions to biologically-based causal agents, diseases are also caused by witches or evil spirits, and so many sorts of amulets, ritual objects, and religious rituals form an intensely practical defense for good health and continued physical survival. I myself would classify religious objects and rituals as "utilitarian" where they pertained directly to the protection of health, food procurement, and fertility for the people who employed them. How to determine this archaeologically is a challenge, of course, but considering the creative ways that archaeologists have managed to address elusive topics of gender, social status, and ideological belief systems, it is certainly not an impossible task.

Thematic Studies in Technology (Continued)

The first pyrotechnologies, other than the relatively low temperatures used in cooking, seem to be used as part of religious rituals which might or might not be classified as utilitarian uses. Such early pyrotechnologies include the roasting of red ochre to brighten its color and the firing of clay figurines, both in the Upper Paleolithic (Schmandt-Besserat 1980; Vandiver, et al. 1989). However, it is not clear that these pyrotechnologies are being used to *create new materials*; that is, they are not necessarily *transformative*. Cooked food is seen as different than raw, but it is not usually described as a new material. (This in itself would be an interesting case for further study.) Similarly, intensifying the color of red ochre might or might not be perceived by its users as the creation of a new material; other than intensification of color, the characteristics of the material are not much changed. More clearly, the firing of clay figurines in Eastern Europe seems to be a part of the ritual activity rather than a desire to create a new material, fired clay. As Vandiver et al. (1989) describe, the clay figurines were formed, then thrown into an existing fire, where they tended to break or explode in the process of firing. The authors interpret this sequence as a case where the *process* of making and firing was itself the desired end—to ensure fertility? good health? bad health to enemies?—rather than the production of a fired clay figurine. Alternately, the disposal of the figurines in the fire might not have been part of the ritual process but rather part of the disposal process, ensuring the proper disposal of these objects. Both alternatives are discussed for other cases of ritual technology in the last section of this chapter. Either of these explanations is better supported by the data than the wish to create figurines from a new material, fired clay. The deliberate firing of clay to create a new material, terracotta, occurs thousands of years after its first use for figurines, in the much later creation of pottery vessels, figurines, and other objects. Another early pyrotechnology, the creation of lime plaster, is also used for primarily religious reasons, in the creation of the plastered figurines and skulls of the Levant (Grissom 2000). In this case, however, lime plaster is also used to make floors at around the same time, and is used to make vessels soon after, both clearly functional (Kingery, et al. 1988). In all of the lime plaster examples, a new material is created—the plaster created from crushed fired rock and clay mixed together into a soft wet slurry that sets to a hard white surface. Fire is necessary to create this material, and the new material is clearly the desired end point of the process of firing and mixing.

Why does it matter if materials are primarily used for "utilitarian" purposes or not—isn't this just quibbling over pointless definitions, definitions that obscure rather than aid our understanding of the topic? To some degree this is very true, and I agree with the points made by Wilk (2001) and others in discussions of how the division between "needs" and "wants" tends to be ethnocentric and moralistic. On the other hand, I have so carefully discussed what I mean by "utilitarian" because I think there are some interesting points to be made about the relationship between the creation of new materials for nonutilitarian purposes and the desire (or need!) for visible methods of status differentiation. Walker (2001: 88) offers a broader explanation as to why this definitional distinction matters for the investigation of ancient technologies. He notes that while it is interesting and useful to know any additional symbolic meanings attached to utilitarian objects, for nonutilitarian objects "what they mean is far more important than what they do." I would twist this slightly, to say that the primary function of "nonutilitarian" objects, what they *primarily* do, is to convey meaning.

Many of the new artificial materials created in the past ten thousand years were first used in the creation of ornaments and amulets, serving vessels, small vessels used to hold precious oils or cosmetics, and other sorts of "display" objects (mirrors, decorative objects, inlays). That is, many new human-created materials were first used for reasons other than physical survival, as luxury or *status-marking items*, items whose primary function was to convey meaning about their owners. This appears to have generally been the case in most regions of the world for the earliest metals, including copper and iron as well as gold and silver. It was assuredly the case for the Eurasian complexes of glazed stones and faiences, glasses, and porcelains. More recently, many of the earliest plastics were used for jewelry and ornamental boxes. Although certainly not always the case, lime plaster and fired clay being equivocal cases and concrete being a clear exception, a frequent general pattern seems to be that when and where these materials were rare, they were used to make ornaments or small display objects. Once the materials became common, with increased production, they were frequently used for tools and other utilitarian purposes as well.

I have observed that most new artificial materials are first employed to make either religious objects or status-marking objects. It is noteworthy that the later cases of the metals and vitreous materials relate to the marking of status differences, while the earlier uses of fired clay and plaster for ritual uses are found prior to the development of strongly hierarchically ranked societies. The example discussed in this section examines the former case in which social status differentiation, the marking of classes, was a primary function of both the new artificial materials and the objects created from them.

STATUS DIFFERENTIATION AND THE DEVELOPMENT OF VITREOUS MATERIALS

The range of vitreous materials created across Eurasia, from glazed stone to glass, provide a variety of examples for discussing the invention and adoption of new materials to create ornaments or other luxury goods that acted as

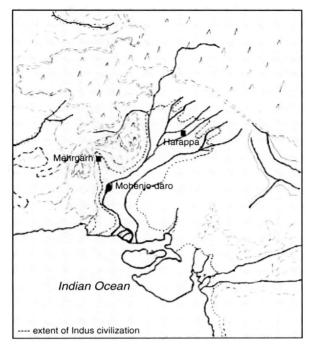

FIGURE 6.1 Map of Indus Valley region, showing sites of Harappa, Mohenjo-daro, and Mehrgarh/Nausharo.

status markers. The manufacturing processes for these materials are described in Chapter 4; here I will look more closely at the cultural context for the development of the talc-faience complex in the Indus Valley region of South Asia, from the seventh through the third millennia BCE (Figures 6.1 and 6.2) Like the other great semi-arid floodplains of the world, in the Indus Valley there is only clay and sand and vegetation for hundreds of kilometers—all other raw materials have to be imported. However, with water, the land is very fertile, and agriculture and animal husbandry have been major occupations and sources of wealth. Furthermore, regions rich in minerals and timber surround the Indus, and in many cases these regions are connected to the Indus Valley by the extensive river transportation systems. Much of this description is typical of all the early floodplain civilizations of the Eastern Hemisphere, the first groups of people to create urban environments and state-level social and political hierarchies. The Indus Integration Era is also the time when powerful leaders in Egypt and the Near East ruled great city-states and empires, so it is not surprising that the same sort of political structure was expected for the Indus. But this does not seem to be the case.

Indus Valley Tradition	Approximate Dates	Equivalent Terms in Other Systems
Early Food Producing Era "Mehrgarh" Phase	>6500 to 5000 BCE	Neolithic
Regionalization Era Balakot Phase Amri Phase Hakra Phase Ravi Phase Kot Diji Phase	5000 to 2600 BCE	Early Harappan or Neolithic/Early Chalcolithic
Integration Era Harappan Phase	2600 to 1900 BCE	Mature Harappan or Chalcolithic/Bronze Age
Localization Era Punjab Phase Jhukar Phase Rangpur Phase	1900 to 1300 BCE	Late Harappan or Chalcolithic/Bronze Age

FIGURE 6.2 Chronological systems for the Indus Valley Tradition, with approximate calibrated radiocarbon dates. (Modified after Shaffer 1992; Kenoyer and Miller 1999.)

The traditional models of centralized state formation are an uneasy fit for the Indus (Kenoyer 1998a, 1998b; Possehl 1998). There is little of the usual archaeological evidence for a ruling elite, either secular or religious: no large temples or palaces, no evidence for a victorious military or an institutional warehousing system, no rich tombs or monumental art. While we do have a few public buildings at Indus sites, like the Great Bath at Mohenjo-daro, the Indus "'monumental architecture" is in many ways the city itself. Rather than an impressive palace complex, we seem to have a city made up of decent neighborhoods, albeit with a range of large to small houses. Long before the Greeks, the people in the Indus region were laying out blocks of housing developments on a rough grid plan, building large-scale sewage and garbage disposal systems, and creating truly massive perimeter walls around their city neighborhoods. Indus art is not monumental but miniature, and it requires a certain level of cultural knowledge to appreciate it, since its value is as much about specialized skill and labor as it is about rare materials. The Indus people shared a cultural style, a weight system, and a script across an area larger than Egypt and Mesopotamia combined, and traded far beyond this area. But there is no obvious evidence for deeply divided social hierarchies, no supreme rulers that we can see. What then was the Indus social and political structure?

A first step in answering this question is to determine how Indus people were marking their status. Clearly some economic and social divisions did exist, but the marking systems are different and apparently more subtle than in other civilizations, as Rissman (1988) suggests through his analysis of urban Indus burials and hoards, and as Possehl (1998) references in his characterization of Indus leadership as "faceless." Kenoyer (1991; 1998b) has discussed the social hierarchies implicit in the types of raw materials used to make red and white beads, and Vidale (2000) and I (H. M.-L. Miller in press-b) have extended this to other bead types. Vidale and I together (Vidale and Miller 2000) played with the idea that the development of new materials over some three millennia was related to the changing nature of social status, especially with the development of cities and complex social and political systems. One approach has thus been to examine the relationship between markers of Indus social relations and the development of new materials, particularly the Indus talc-faience complex.

All three of the earliest urban societies of the Eastern Hemisphere, Egypt, Mesopotamia, and the Indus, developed complexes of vitreous materials that include glazed stone and a group of glossy, silica-based materials most often called "faience," as described in Chapter 4. The development of these vitreous materials in these three regions probably represents technological stimulus and diffusion, with each region aware to some extent of the materials developed in the other regions, but manufacturing their own objects. There are overlaps in the production processes and types of materials in the three regions, but each region seems also to have made its own innovations and followed its own path of development. For example, only in the Indus was a type of faience developed that included fragments of talc as well as silica in the body.

The talc-faience complex of materials well represents the long development of Indus artificial materials, beginning for this complex with the heat treatment of talcose stone in the sixth millennium (after 6000 BCE). A remarkable property of talc (also called steatite) is that although it is soft and multicolored in its natural state, when heated to high temperatures (above 1000°C), all types of talc become hard and many become bright white, even some black talcs. This striking material transformation may have given talc/steatite a special significance for the Indus. Vidale (2000) has speculated, on the basis of his ethnoarchaeological work in Baluchistan (Vidale and Shar 1991), that the importance of talc in the Indus bead assemblage may in part be related to its startling transformation from various colors to a bright white after firing. This color change may have served as a material illustration or symbol of religious beliefs. So it is noteworthy that beginning around 6000 BCE in the burials at the earliest Indus Valley Tradition site of Mehrgarh, there is an increase in fired talcose beads, which are white in color, in parallel with the decrease in white shell beads. By the start of the Indus Integration Era around 2600 BCE, shell beads are relatively rare in both burial and non-burial contexts at numerous Indus sites (Barthélémy de Saizieu and Bouquillon 1994; Kenoyer 1995; Vidale 1995; H. M.-L. Miller in press b). This rarity has nothing to do with difficulty of access to the raw material, as shell bangles and other shell objects are quite common. In fact, shell is still the primary material used for bangles for Harappan Phase Integration Era burials (Kenoyer 1998a: 144), but bead ornaments are primarily made from talc in various forms. Thus, shell continues to hold ritual value in some forms, but fired talc beads replace shell beads altogether in all contexts. Again, this valuation of talc may have to do with its transformative quality, something not possible with shell.

Our knowledge of the talc-faience complex bead materials prior to the Indus Integration Era is based primarily on work done by the French archaeological project at Mehrgarh and Nausharo (Barthélémy de Saizieu and Bouquillon 1994, 1997; Bouquillon and Barthélémy de Saizieu 1995; Barthélémy de Saizieu 2004). The first talc/steatite beads found at Mehrgarh are unfired, usually with a natural color of black or dark brown, and alternated with white shell beads. In the sixth millennium, white-fired talcose beads begin to appear (Figure 6.3). By the beginning of the fourth millennium, more than 90% of the talc beads from Mehrgarh were fired white, and the first blue-green silicate glazes on talc beads are also found at this time. Just prior to and at the beginning of the "Pre-Indus" periods at Nausharo, from about 3200 BCE onwards, there is an increasing predominance of discoid forms and tiny sizes in fired talc beads, and a number of new materials appear, including talcose-faience, siliceous faiences, and possibly talc paste. These materials continue to be used throughout the Indus Integration Era (2600–1900 BCE) and have been common finds at most Indus sites (discussion and references in Kenoyer 1991; Vidale 1992, 2000). Blue-green glazed talcose stone was used exclusively for beads, while white-surfaced fired talc was used to make the most common inscribed Indus materials, seals, tokens, and tablets. The Indus microbeads, only one millimeter in diameter and length, may have been either individually cut and ground from talcose stone or produced from a still undefined sintered talc paste mixture, then fired. Talcose-faience, a material with talcose fragments in a sintered silicate matrix, may primarily be a transitional material employed in the first periods of faience manufacture, but the very small number of tests for materials dating to the Indus Integration Era makes this an entirely open question to date (see Chapter 4, Vitreous Silicates section). It may equally represent a material used for particular purposes, and/or of particular symbolism. Siliceous faience, which turns quartz sand or ground pebbles and a little copper dust into a brightly glazed blue-green sintered silicate object, was widely used for bangles, beads and other ornaments, inscribed tablets, inlay pieces, small vessels, and small figurines or amulets. Classification of the exact material used to make a particular object is difficult, as they are almost identical in appearance even under low magnification, and descriptions in the literature are thus often incomplete or confusing; see Miller (in press-a) for a detailed, descriptive terminology for the various materials in the Indus talc-faience complex. These artificial materials are linked not only by their very similar physical appearances, but also in their overlapping raw material components.

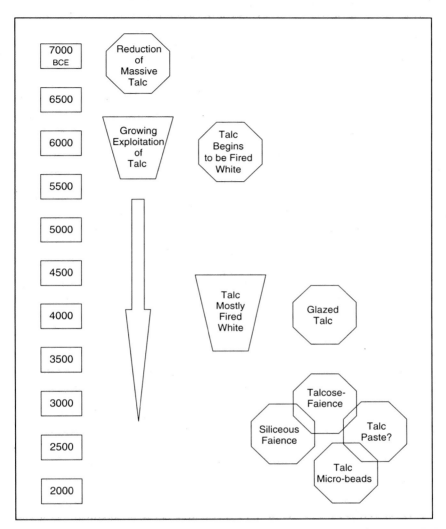

FIGURE 6.3 Development of new talcose- and silicate-based materials (talc-faience complex) in the Indus Valley Tradition. (Data primarily from Mehrgarh-Nausharo studies by Barthélémy de Saizieu and Bouquillon—see text.)

In addition, they likely had connections during the process of production, whether through the recycling of by-products such as talc powder or in the use of similar techniques of production. These diverse talc-faience materials may have been made in the same workshops.

All of these vitreous materials require more skill and more specialized knowledge to produce than beads cut from massive talcose stones, particularly with the need to perfect paste and glaze compositions and firing techniques. The new materials also allowed the use of talc and/or quartz powder, and so the use of waste materials and lower quality talc. This explosion of new materials is thus a brilliant example of the technical virtuosity of the Indus craftspeople, and also provides insights into Indus society (Vidale and Miller 2000). "Indus technical virtuosity," as we defined it, refers to the striking Indus characteristic of inventing, adopting, modifying, and diffusing complex techniques across the large Indus region. These techniques were not used to create monumental objects or large symbols of religious or secular power, but were used primarily for the production of small ornamental objects, objects that were worn or otherwise displayed. In our opinion, the small size of these ornamental objects did not preclude their social importance, particularly for communicating social roles, and the development and encouragement of Indus technical virtuosity reflects and is a reflection of strategies of social patterning from the fourth through second millennia BCE, or even earlier.

The creation of these many new artificial materials occurs around the time of the development of the urban, heterogeneous Indus Integration Era social and political structure. There seem to have been a number of levels of social status created at this time, rather than just a bipartite division between "commoners" and "elites." Looking more broadly, this seems a characteristic not only of the Indus, but of many of the Western Asian civilizations of the third and second millennia BCE. This extended system of social levels would need new methods of marking or signaling these varying levels of status. One such expanded status marking system has been suggested for various prehistoric and historic periods, where an increased use of artificial materials is tied to a widening demand for status or luxury goods, with the development of a middle-level elite, a bureaucracy, and/or a wealthy urban class (e.g. McCray 1998; Moorey 1994: 169; Vidale 2000; Vidale and Miller 2000; H. M.-L. Miller in press-b). These new materials could be employed to create status symbols for such middle-level classes, allowing an extended hierarchy of status in societies ranked with increasingly complexity. How can we determine if the case of the Indus talc-faience materials represents a similar situation, where new materials are being used to mark an expanding number of social levels? The first step is to determine the relative value of materials and objects, old and new, natural and artificial, to create a ranking of the relative value of materials.

Determining Relative Value

The relative values of materials and objects is culturally-specific, as researchers have discussed for a number of situations (e.g., Helms 1993; Lesure 1999).

This makes it a challenge to identify scales of value in the past, particularly when we have few historic or oral records as a guide. Furthermore, the economic value of materials and objects may be very different from ritual or status scales of value within the same society; an object of low economic value may have quite high ritual value. For example, Solometo (2000) discusses the relative value of materials used by the Hopi to make pigments for ritual wall paintings, as mentioned in the section on Ritual Technology at the end of this chapter. For this case, exotic rare materials were highly valued as pigments, but the value of other objects relied on other associations besides scarcity, such as color or association with ancestral dwelling places. To deal with these complexities, archaeologists must employ a large number of diverse data sets to assess the relative value of objects and materials, including textual and pictorial materials, oral traditions, historical and ethnographic analogies, the archaeological contexts and archaeological rarity of finished objects, and evidence for their curation (Figure 6.4). The ambiguity of many of these sources of data for the Indus is discussed and assessed in Miller (in press-b).

Technological considerations are also used to assess value, such as the *estimated time of production*, often used as a proxy for labor costs. For the Indus, estimated time of production has been approached through a number of ethnoarchaeological and experimental projects focused on Indus materials, such as the stone bead studies discussed in Chapter 3 (e.g., Kenoyer, *et al.* 1991, 1994; Roux, *et al.* 1995; Roux and Matarasso 1999; Roux 2000; Vanzetti and Vidale 1994; Vidale 1995, 2000). The basic idea is that only wealthy elite can afford the costs involved in craftspeople producing few artifacts over a long period of time, the labor expense of the object. For example, the lengthy time involved in drilling hard stone beads, such as agate or carnelian, would theoretically make them more valuable than softer stone beads that required less time to create and so less labor cost. However, rather than simply calculating the times spent on each production stage to measure expense, most of the Indus studies have placed equal emphasis on specialized skill and/or knowledge, as is further discussed below. Many studies, well summarized by Underhill (2002: 6–8), have focused on such status markers created with significant labor investment, particularly in socially and politically ranked or middle-range ("chiefdom") societies.

Texts & Pictures	Archaeological Context	Time of Production (Labor Costs)
Oral Traditions	Archaeological Rarity	Difficulty of Procurement
Ethnographic Analogies	Curation	– Access to Materials
		– Complexity of Production

FIGURE 6.4 Data sets used by archaeologists for the assessment of value.

Another technological consideration used in the assessment of relative value is the *estimated difficulty of procurement* (Figure 6.4, far right). This category includes both the procurement of raw materials from which the objects are made, and the complexity of production, involving specialized knowledge, skills, and tools. Many archaeological models of the relationship between status and display or "prestige" goods focus on the former, primarily the procurement of exotic raw materials (e.g., Helms 1993; Bellina 2003), as exotic raw materials are often common component of status items. Both for status based on intensive labor investment and status based on exotic raw materials, ornament styles are a particularly useful method of marking social information through display (Wobst 1977). For the Indus, it is especially important to also include complexity of production in any assessment of status markers, given the many complex craft traditions found in the Indus with the creation of new artificial materials such as faience, fired talc, metal alloys, and stoneware. The timing of the development of these new materials is very suggestive, occurring at the same time as there seem to be increasing numbers and types of social classes.

Several recent Indus studies have focused on these two technologically-related data sets for assessing relative value, primarily on both aspects of the relative difficulty of procurement but also indirectly on labor costs. Figure 6.5 is a diagram based on earlier charts used by Kenoyer (1992) to represent relative types of control over different craft industries, and charts used by Vidale (1992) to assess the relative value of object types made from different materials. In this *expected relative value* diagram, the relative accessibility of the raw materials needed to produce a given type of object is assessed and graphed along the horizontal (x) axis, while the vertical (y) axis represents the assessment of the relative complexity of production. Neither assessment is a straightforward procedure. For example, the assessment of the difficulty of accessing raw materials must allow not only for the physical distance to sources, but also the environmental conditions and the social situations that affect the ease with which craft producers can obtain these materials. A metal ore source accessible by river transport may be more accessible to producers than a nearer source accessible only by walking. Or a lithic or mineral source on one end of a seasonal pastoral round may be more accessible to producers on the other end of the round if they have trading relations with the pastoralists. For the vertical (y) axis, the relative complexity of the production sequence, we need to have some measure of the degree of specialized knowledge needed to produce certain objects, not to mention the probable restriction of such knowledge, in order to estimate the difficulty of production for each type of object. The information needed to rate the relative complexity of production involves *at minimum* the construction of production process sequences for each object, as described in Chapters 3 and 4, detailing

Thematic Studies in Technology (*Continued*) 215

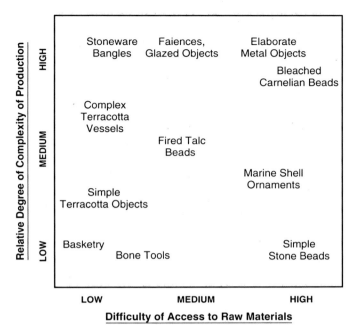

FIGURE 6.5 Expected relative value diagram; this example uses Indus Integration Era object/material types.

the number of production steps and tools, and an indirect or direct estimate of the time needed for production. But this is not enough to rate the relative complexity of production; an assessment of the technical knowledge involved may be even more important (e.g., Wright 1991).

Thus, diagrams like Figure 6.5 approximate the *expected relative value* of objects, as they explicitly combine estimations of the *relative* difficulty of access to materials and estimations of the *relative* complexity of production, the later usually also implicitly including labor costs. All of these variables are extremely difficult to measure archaeologically. These diagrams are further simplified in that the type of object being produced is not specified, and not all objects follow the same production sequence even if they are created from the same raw materials. However, such diagrams can be used to assess the expected relative values of objects on a broad scale; see Vidale and Miller (2000) for an example. These diagrams are especially valuable in that they allow the assignment of expected relative value to objects produced by one craft versus another. For example, clay suitable for the production of all types of fired clay objects is universally available in the Indus floodplains. Deposits of stone and metal ore are much more restricted in distribution, to the edges of the Indus region. Strictly in terms of raw material access, objects of clay should thus have been less valuable than objects of talc or agate or copper. However, when the relative value imparted by the complexity of producing an object from these raw materials is added, relative values may change. For the fired clay crafts, complexity of production ranges from the simple techniques used to make and fire terracotta "cakes"; to the more uniformly processed and fired terracotta bangles and figurines; to wheel-thrown and elaborately painted pottery; to the extremely complex firing regimes of stoneware bangle production, so called because they were fired to the point of sintering and break like flint or obsidian. All of these objects are made with the same raw materials; any difference in expected relative value lies in the complexity of their production.

In sum, the schematic organization of data shown in Figure 6.5 simplifies the depiction of the very complex data involved, much as graphs and statistical test results are used to simplify the portrayal of complex numerical data. As noted, the diagram is particularly useful as it allows the comparison of expected relative value for objects produced by different crafts. These diagrams can also be used to examine changes in value for material culture assemblages through time (Vidale and Miller 2000). Furthermore, these production-based assessments of the expected relative values of objects (Figure 6.4, column 3) can be compared with relative value assessments based on archaeological data such as context, rarity and curation (Figure 6.4, column 2), and/or value assessments based on ethnographic parallels or historical documents (Figure 6.4, column 1). These comparisons between value assessments based on different types of data set up a system of cross-checks for the difficult task of appraising ancient values.

For example, Indus Integration Era stoneware bangles, produced from a widely accessible material but with a very complex technology, were higher in value than copper bangles, which were of imported materials but generally produced in a relatively simple manner. Such an assessment is supported not only by these *expected* relative values, but also by the archaeological evidence on context and rarity. Stoneware bangles have a very restricted distribution, being found only at the largest sites, an unusual restriction of consumption for the Indus. They also had a very restricted production, as they were produced only at Mohenjo-daro and Harappa, and possibly one other site (Blackman and Vidale 1992). In addition, most or perhaps all of these stoneware bangles had individual inscriptions. Furthermore, the stoneware bangles provide sound evidence that they were valued for the material itself, rather than the finished appearance of the object. In spite of their very careful and elaborate manufacturing process, there is little attention to their surface finishing. Breakage scars left on some bangles from sticking during firing were only roughly ground down, leaving them quite visible (Vidale 1990). While many or most were inscribed, the inscriptions are roughly scratched, in great contrast to the highly polished finish left by the manufacturing process (Kenoyer 1994). The value in these objects seems to have been based on the elaborate production required to create this material, rather than on details of perfect finish—the type of skill valued may be mastery of the deep transformation of clay to "stone," not perfection of surface. For the Indus, it is also possible that a portion of the value given to stoneware bangles, like talc and faience objects, lies in their transformation by fire to a new artificial material, as I will further discuss below.

For the specific case of Indus talc-faience materials, Vidale and I (Vidale and Miller 2000) placed them as objects made from materials not so difficult to access, but requiring quite high levels of technological elaboration (Figure 6.5). These materials range from low to medium difficulty of access, as the talcose stone and quartz minerals needed were located in several regions around the immediate edges of the Indus Valley, so that they could likely be procured in sufficient amounts in a steady supply through reliable trading connections (Law 2005, 2006). But all of these materials involved production processes of many steps and a great deal of specialized knowledge, resulting in a high rating for technological elaboration. This combination, we thought, made them ideal status markers for the growing middle levels of status in the urbanizing Indus civilization. These ornaments and other display items, made of heat-transformed talcs and/or various faiences, could function as symbols of status for the growing ranks of merchants, workshop owners, and bureaucrats in trade, craft production, and urban management during the Indus Integration Era (Vidale 2000; Vidale and Miller 2000). Moorey (1994: 169) has similarly suggested that Mesopotamian faience and glass production was stimulated by "widening social demand, rather than depredation in the richer sections of society."

SOCIAL RELATIONS AND THE RELATIVE VALUE OF INDUS TALC-FAIENCE MATERIALS

Was this suggestion correct? What was the relationship between the demand for new artificial materials and the increasing diversification of social and economic classes in the Indus? This is a point that turned out to be rather involved but highly revealing for the Indus, and is especially interesting in light of some unusual aspects of Indus material use in comparison to the contemporaneous early civilizations of the Near East and Egypt, particularly the relative absence of lapis lazuli and the prevalence of talc/steatite. As outlined below, for some cases, such as red with white colored beads, the proposal that new artificial materials were employed to fill a need for more middle-level markers of status in an expanding social hierarchy is supported. However, in other cases, particularly for beads and other objects made of white heated talc, the primary use of artificial materials is not as markers of *social hierarchy*, but rather as markers of *social unity*. That is, the use of new materials as social markers is not limited to a marking of social hierarchy, but also represents culturally-specific desires and values associated with the "Indus" way of life. Some of these culturally-specific values may relate to Indus-wide belief systems, including a high esteem for heat-transformed materials and the desire for specific colors and high shine.

Such new markers of cultural unity would be needed in the increasingly heterogeneous societies of this period, where the problem of uniting disparate cultural and social groups into larger-scale, state-level political unit(s) had to be addressed. The need to promote a feeling of social and political unity would be a problem for the stress of urban proximity as well as for the danger of dissolving ties over large distances. On reflection, it is apparent that creating new markers of social unity would be as much of a pressing need for emerging state-level societies as the creation of markers of social hierarchy. Thus, with closer study, there are a number of reasons why the development of artificial materials as markers of social meaning may have been encouraged in the Indus.

In 1991, Kenoyer suggested that there were numerous imitations of natural red with white carnelian beads, with imitations occurring in multiple materials while preserving the same form and appearance (Figure 6.6). These materials include artificially produced white etchings on "natural" red carnelian, red and white banded faience, two kinds of red-painted white talc/steatite, and even white-painted red terracotta. Kenoyer (1991) suggested that these various imitations represented hierarchies of value and thus hierarchies of socioeconomic status. The naturally marked agate beads were of highest value and worn by high status individuals, and the least rare and/or technically complex beads (the terracottas) of lowest value and worn by low status individuals. He expanded on this discussion in his article about Indus wealth and socioeconomic status (Kenoyer 1998b). Based on Kenoyer's arguments, Vidale and I (2000) then suggested in 1997 that the many new talc-faience complex materials produced in the Indus region during the third millennium BCE were employed to create status markers for an expanding middle level bureaucracy and socioeconomic elite, along with other objects of low access cost but high technological complexity. This suggestion that new artificial materials of middle overall value were created to provide imitative luxury goods for new middle levels of status, while the highest elite continued to acquire rare natural materials, fits perfectly with the data for red with white beads. But when I subsequently began to look more closely at the data for

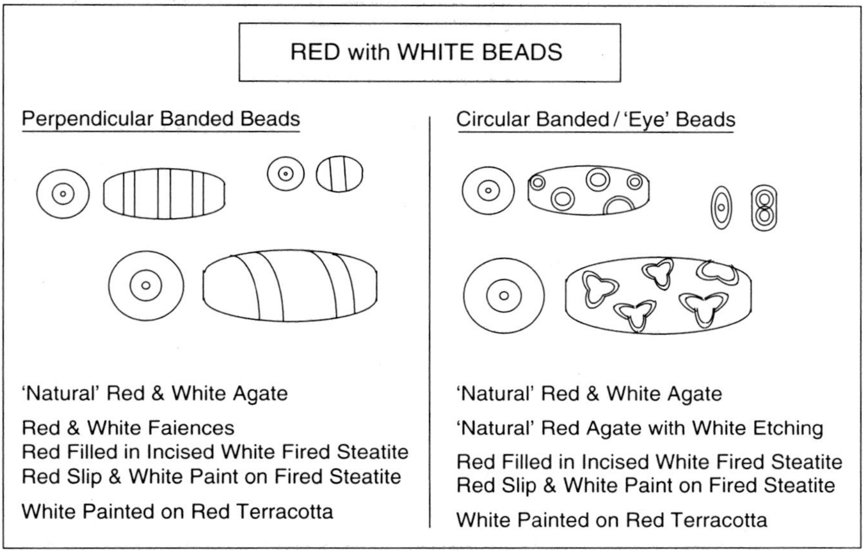

Thematic Studies in Technology (*Continued*) 219

FIGURE 6.6 Materials used to make red with white beads in the Indus Integration Era. ("Mature Harappan period")

beads made of talcs and faiences, the majority of which were white or blue, this hierarchy of value no longer seemed so clear.

Unlike the red bead assemblages, exotic natural stones of white and blue, such as quartz, white agate, ivory, alabaster, lapis, and to some extent turquoise, were *not* used to make highly valued beads (Figure 6.7). This is not a problem of lack of access to the raw materials; these materials were available to Indus people and at least the white materials were often used to make other objects. Instead, white and blue beads made of artificially created materials seem to have been the most highly desired, particularly materials altered or created by very high heat, such as fired talc, talc/steatite microbeads, and the various faiences. These are unlikely to be imitations of rare natural materials, because natural exotics were simply not used, with the possible exception of turquoise, which is found in small quantities and used only to make beads, not seals or other display objects. The blue faiences and glazed materials of the Indus Integration Era are typically light blue to bright, slightly greenish blue, or even "apple-green." (This last, rare color of faience could have been an imitation of green agates or other stones; see (H. M.-L. Miller in press-b). It is possible that blue-green glazes may be imitations of turquoise, but the turquoise found at Indus sites does not readily support a high valuing of turquoise across the Indus social system. It is not of very high quality nor as plentiful as one would expect, given its availability in nearby Baluchistan and

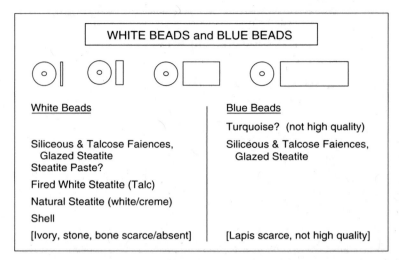

FIGURE 6.7 Materials used to make white and blue beads in the Indus Integration Era. ("Mature Harappan period")

the ease of working this softer stone. It may have been valued by a subgroup within the Indus civilization, another complication to any determination of value. In any case, the light blue to sky blue glazes, the most common blue colors, are not imitations of any natural stone at all, as no natural materials of these colors are found in the Indus bead assemblages. Most significantly, we find no dark blue glazes or faiences at all during the Indus period; that is, no potential imitations of lapis lazuli. This absence or underutilization of rare exotic stones is particularly surprising for lapis lazuli, which was so highly valued in contemporaneous Egypt and Mesopotamia. Lapis was available to the Indus people through their trading networks, and their presence in Baluchistan and Afghanistan, most notably the site of Shortugai in the Oxus region (Francfort 1989), has been assumed to indicate their interest in acquiring lapis. However, early excavators with experience at Mesopotamian and Egyptian sites of the same period mention the surprising lack of lapis at Indus sites (e.g., Mackay 1938: 499). What little lapis has been found even within the cities of the Indus Valley is often of relatively poor quality. Significantly, at least at Harappa itself, it is not until the very end of the Indus Integration Era that we see the development of cobalt and dark blue, lapis-colored faience (Kenoyer 1998a: 176).

Thus, for the white and the blue beads of the Indus Integration Era, there is no hierarchy of value with rare materials at the top, representing an equivalent socioeconomic hierarchy for their possessors. In addition, in spite of other evidence for their apparently high value such as their association with other high-value materials, their context, and their technological complexity (H. M.-L. Miller in press-b), beads made from talc-faience materials were found in large numbers and were widely distributed across sites. They are thus unlikely to represent high- or even middle-level status markers. This is particularly the case for the white fired talc beads, although their large numbers partially reflect the large numbers of these small beads needed to make up a single ornament such as a necklace or hair ornament. These conclusions clearly negate part of the argument I made with Vidale, about these new pyrotechnological materials serving primarily to make luxury goods for middle-level elites, while high-level elites continue to use objects made of rare natural materials. This data also negates the assumption that these new talc-faience artificial materials were desired and valuable, but of lower value than rare natural materials. Clearly, this ranking of artificial vs. rare natural materials *can* be the case, as seen in the red with white beads, and perhaps the apple-green beads, but is not *necessarily* the case, as seen in the white beads and in some of the blue beads. In other words, there is more than one type of culturally-specific value system underlying the relative values of Indus materials, even within this single artifact class of beads. In some cases, artificial materials are imitating rare natural materials in a hierarchy of value, while in other cases, artificial materials are highly valued in themselves, and do not appear to convey information about hierarchical status.

These white and blue beads might represent another sort of social information, however. It is noteworthy that while this explosion of new material types is taking place, the forms of these beads are similar or even identical in all the various materials. The majority of Indus Valley Tradition talc and faience beads were simple discs or cylinders with no surface decoration, and change very little over more than a thousand years. We know from other objects that the Indus craftspeople had the necessary skill to create very elaborately formed objects. This uniformity of form is thus a deliberate choice during the Indus Integration Era, a choice that I argue was made to emphasize social unity and an allegiance to the Indus ideological system, while the variety of rare and artificial materials employed for some of the same beads could be used to emphasize a hierarchy of economic or social status (H. M.-L. Miller in press-b). Elsewhere, I have examined the range of social messages in the development and adoption of new artificial materials by speculating on the relative values of other sorts of new materials for the Indus, such as the range of metal types and the range of fired clay materials, especially Indus stoneware (H. M.-L. Miller in press-b) . Here, I instead continue to focus on sintered talcose and siliceous materials, and examine their possible use for other types of objects for conveying information about various sorts of social relationships.

The vast majority of talc-faience complex objects are beads; indeed some of the materials in this complex may have only been used to make beads, such as talc paste (if it existed). The second most common use of the faiences, a use unique to the Indus, was to make circular, closed bracelets referred to as bangles. A particular type of faience used only in the Indus may have been specially devised for the physical stresses associated with the bangle shape (McCarthy and Vandiver 1991). Unfortunately, very few studies have been done of Indus bangles, although they were produced in great numbers from a range of clay-based materials, faiences, metals, and shell, and represent one of the largest artifact classes found on Indus sites (Kenoyer 1991, 1998b; Thomas 1986). Nevertheless, there are indications that the assemblages of Indus bangles, like the beads, represent multiple value systems and make multiple social statements. There are bangles of similar shape and decoration in a variety of materials, such as the peaked cross-section bangles with a chevron design made of shell, faience, and possibly other materials (Kenoyer 1998b: Figure 8.11). However, the determination of any hierarchy of socioeconomic value is complicated by the fact that shell bangles clearly had a special ritual value, as shell bangles are the only type of bangle found in Indus burials although they are not nearly so dominant in other contexts (Kenoyer 1998a: 144). Rissman (1988) discusses the special case of ritual burial value, as a separate social statement apart from wealth-based hierarchical systems, through his assessment of Indus horde and burial contexts. Although he does not examine bangles specifically, his general argument compliments the idea that this association of shell bangles with burial contexts may represent the use of shell as an ideological or social marker that is not related to socioeconomic status. There are clearer hierarchies of value in the copies of the black stoneware bangles made in fine clays fired both red and black, as well as in common red terracotta, all identifiable by their distinctive flattened tear-drop cross-section (Kenoyer 1998b: Figure 8.11). Unlike the red with white beads, however, red faience is not part of this hierarchy, and there is no natural stone material at the top of the hierarchy. Here again, the stoneware bangles are not imitations of exotic stone bangles, but are the most highly valued type of bangle themselves. For other bangle types, only artificial materials were used (fired clays, metals, faiences). Metal bangles were typically plain circlets, but faience and terracotta bangles were made in a wide variety of shapes, some unique to the blue faiences (Kenoyer 1998b: Figure 8.11). Further study of the Indus bangles for clues to social and ideological statements would be of great interest.

During the Indus Integration Era, faience was also used to make small two-sided, three-sided, and four-sided tablets impressed with Indus script and scenes. The only other material used for impressed tablets was terracotta, although somewhat similar two-sided incised tablets with images and script made of copper are found only at Mohenjo-daro (Kenoyer 1998a: 74). Another major use of faience materials was to make small vessels, thought to be used for holding special oils, perfumes, or other expensive materials requiring impermeable vessels. These small, delicate vessels were typically made in blue or white faience. Surprisingly, talc and other stones are never used to make vessels, so there is no indication of hierarchical imitation in this use of faience. Faience was also used to make small figurines, which were made of stone and metal as well, and to make inlay pieces for jewelry or possibly furniture, which were also made from shell and stone. Unfortunately, there has not been enough research into the figurines and inlay to say anything about possible value associations.

The uses of fired talc/steatite stone, the remaining material in the talc-faience complex, are fewer but perhaps more intriguing. Fired white talc/steatite beads are by far the most common type of bead found at Indus sites (Kenoyer 1998b: Figure 8.15). After bead production, the most common use of fired talc was to make the characteristic Indus stamp seals and small tablets or tokens inscribed with lines of Indus script. The inscribed tablets were only made from talc, although a few copper objects of similar type with raised script letters have been found at Harappa (Kenoyer 1998a: 74). Similarly, the stamp seals were only rarely made from any other materials, such as metal. These exceptions may have been idiosyncratic personal choices, or may represent a very truncated hierarchy pyramid; examination of the material type used for seals in comparison with script characters, imagery, and location of find might be revealing. The dominance of talcose stone for seal production is not a functional choice related to their use, as stamp and cylinder seals in other contemporaneous civilizations are made from a much wider variety of stones and other materials.

Even at this stage of investigation, it is striking that the vast majority of objects whose primary function was to convey inscribed information were made from materials in the talc-faience complex. (The Indus script was also frequently found on terracotta vessels, metal tools, and some other objects, but the primary function of these objects were as containers, tools, etc.—the script notations are secondary.) The focus on talc-faience materials looks like a deliberate choice based on a cultural value. This is particularly true for the fired talc stamp seals. The Indus focus on a single, relatively common material for the vast majority of the square stamp seals produced over some seven centuries is in contrast to contemporaneous Mesopotamia, where cylinder seals are made in a range of materials of varying value and changing popularity through time (Collon 1990). The material of the stamp seals, then, seems unlikely to be a sign of status, as is also true for the dominance of fired talc in the Indus Integration Era bead assemblage. Instead, the use of this material seems to relate to other issues, and a sense of an "Indus" identity is the most likely explanation. H. C. Beck christened the Indus a "steatite society" in the 1920s, and data since then has only enhanced this reputation (Vidale 1989, 2000).

Furthermore, the *value* of fired talc must come from attributes other than imitation of rare natural materials. As noted above, when talc is heated to high temperatures (above 1000°C), it becomes hard and bright white. This material transformation may have given talc/steatite a special significance for the Indus, as Vidale (2000) has speculated, and this color change may have served as a material illustration or symbol of ideological beliefs (see section above on Status Differentiation). Other valued new materials were also transformed by heat. Faience production turned quartz, plant ash, and copper dust into a brightly glazed blue object. Stoneware bangle production transformed tan-colored clay to a metallic black chert-like material. Even the case of the red with white beads partially supports the importance of color and material transformation through heat, for while natural red agates were the highest valued, these "natural" agates were usually heat treated (albeit at much lower temperatures) to redden the color and enhance the chipping ease (Kenoyer, *et al.* 1994; Roux and Matarasso 1999). However, I want stress that the Indus regard seems to have been for the transformation process, not for fire itself. There are no clear symbols or scenes of the veneration of fire in the Indus cultural material. The so-called "fire-altars" of Kalibangan are identical to well-documented cooking hearths and pottery-production kilns found at other sites (H. M.-L. Miller 1997: 45–46; H.M.-L. Miller 1999: 45). As a pyrotechnologist, I would like to believe the Indus people valued fire and its uses, but certainly there is no archaeological evidence to show that they held it in religious veneration, beyond the likely veneration of the home hearth typical of most societies in the past to varying degrees.

A regard for color transformation alone does not provide a complete explanation; note that most of the materials above also became harder, and sometimes lustrous. Shell can be calcined, and so whited by heat, but this weakens the material rather than hardens it, and calcining also destroys the reflectivity (shine) of the shell surface, making it matte. Kenoyer (1998a: 96) has pointing out the high reflectivity or shine of the faience materials and their lack of discoloration in contact with body oils, unlike untreated talc/steatite, ivory, bone, some types of shell, turquoise and lapis lazuli. Agates and other crystalline stones, which were extensively used and valued by the Indus people, can actually increase in shine with exposure to body oil. The relative indifference of Indus people to lapis and even turquoise can thus be explained by the Indus esteem for materials transformable in color and nature (hardness) by heat, and for materials with high reflectivity.

In sum, one single attribute cannot explain all cases—all of these value systems are interwoven in complex ways. The lack of natural white and blue materials with the attributes valued by the Indus people, such as transformation of color and nature by heat as well as high reflectivity, explains the truncated pyramid of value for white and some blue beads. Thus, the apparently strange lack of interest by ancient Indus people in one of the most valued materials of their age—lapis lazuli—makes sense when viewed from the perspective of their own cultural value systems, particularly the value placed on materials transformed by heat, and materials with high reflectivity. And characteristically, when the Indus craftspeople could not find natural materials with the attributes they desired, they simply created new materials. The Indus people are noteworthy for their cultural expression of "[n]ot the power of conquering, but rather the power of creating; from the abstract universes created in their urban organization to the artificial stones of their microbeads" (Vidale 1989: 180). Technological virtuosity indeed.

Artificial Materials and Cultural Value Systems

Overall, I found throughout my research into Indus pyrotechnologies that Indus craftspeople were extremely technologically innovative in the creation of new materials. This was particularly true for the sintered talcose and siliceous materials, where we see a virtual explosion of new materials during the third millennium BCE, at the same time as the development of the Indus civilization during the Indus Integration Era (Figure 6.3). In some cases, artificial materials do seem to have been used to make imitations of rare natural materials, apparently to increase the number of status markers in an expanding socioeconomic hierarchy. But in other cases, the relative value hierarchies yielded some surprising results, with artificial materials more highly valued than expected, and not imitating rare natural materials at all. The pyramid of value is truncated for many of the new materials, particularly the talc-faience complex and stoneware, with no rare raw materials used by a high elite. Culturally-specific desires and values were more complex than simply a high valuing of exotic rare materials; a number of desired attributes, in some cases overlapping, relate to this development of new materials. Rare natural materials were indeed desired and imitated. But materials transformed in color and nature by heat seem to also have been highly valued, as were materials of high reflectivity. Any or all of these categories may be related to Indus ideological systems.

In short, there were additional reasons why the development of new artificial materials for display items may have been encouraged in the Indus, beyond their role as a marker of status for new middle-level socioeconomic classes. Particularly for beads and other objects made of white heated talc, the primary use of many new artificial materials seems to be not as markers of social hierarchy, but rather as markers of social unity. The Indus Integration

Era, like earlier periods in Mesopotamia, marks a time when new social and political configurations were appearing for the first time. Humans moved into the dense social landscapes of cities, and formed ever-widening social relationships. Larger and more complex political structures were needed to deal with the tensions and complexities required by the increasing heterogeneity of social groupings. New markers of social relationships were thus needed, both to clearly show the developing social and political hierarchies, and to signal general membership in these new social and political organizations. So new symbols of belonging were needed, and new artificial materials offered opportunities as new markers of social and ideological meaning.

Although apparently only the Indus people so highly valued talc/steatite, siliceous vitreous materials such as faience were widely distributed in the third and second millennia BCE, from Asia to Europe. The uses and values of vitreous materials in each of these societies is somewhat different, as one would expect. In all, however, vitreous materials were used to make luxury, status, or display items—ornaments, small figurines, small vessels for expensive reagents—but for a much wider range of society than the top elite. In later periods, glass comes to play a similar role (Fleming 1999; McCray 1998; Vidale and Miller 2000), then porcelain (Kingery 1986; d'Albis 1985), with the two latter also being used for display-oriented serving vessels. As was clear in Chapter 4, a wide variety of overlapping vitreous materials were developed in various regions and periods. The very tangled nature of this group of pyrotechnologies in itself offers a wealth of information about the reasons for innovation and the creation of new materials. Once we have a better idea of the appearance, disappearance, relative proportions, and contexts of these various materials in different regions through time, we can begin to look for patterns. It will be enlightening to see if increasing demand for such luxury, status, or display objects across a wider range of economic, social, and political classes frequently encourages the development of new imitations made from cheaper, more accessible raw materials. As Wilk (2001) discusses at length, understanding "needs" and consumption choices is essential to investigating the history (and prehistory) of technology. Furthermore, it is clear from the cases discussed in Chapter 5 that adoption of new inventions can be based as much on social or political conditions, or cultural beliefs, as on efficiency and practical use.

TECHNOLOGIES OF RELIGIOUS RITUAL IN THE AMERICAN SOUTHWEST

> A Hopi kiva is not only a place for the expression of religious beliefs; it is also a machine used to bring harmony to the world.
>
> (Walker 2001: 87)

I have said very little about the role of ritual within the practice of technology, although ideationally-based ritual actions and choices are an integral part of many technological systems, such as the African iron production systems summarized in Childs and Killick (1993), or the use of particular materials as examined in Hosler's (1994a; 1994b) work on metal use and production in ancient Mexico. However, this section is not about the role of ritual in either production or in technological style, but rather the analysis of ritual practices and objects from a technological perspective. Rituals, religious and otherwise, can be analyzed as technologies. This is just one of many possible approaches to the archaeological study of ritual, and it is not common, but there have been some interesting insights from taking this perspective, as seen in the works discussed here by Solometo and Walker for the American Southwest. In these studies, data from ethnohistorical sources and archaeological finds are applied to archaeological questions. Solometo (1999; 2000) investigates the production process of religious paintings in historic Pueblo rituals, while Walker (2001) is interested in the archaeological identification of religious ritual. Both examine religious ritual within the context of an overall technological system, as they are interested in the social and cultural context in which these rituals are occurring. These complementary studies show the range of ways technological approaches can be applied to the study of religious ritual.

Rituals can be defined as stereotyped patterned behavior or activities of any kind, whether religious, political, social, or a mixture. For example, worship services are a type of religious ritual, while everyday greeting exchanges are social rituals. Ceremonies of installation into office are political rituals, but if a leader can take on supernatural traits, associated rituals are both political and religious. In fact, rituals most commonly incorporate a mixture of goals, as in most aspects of life; attendance at a worship service may be undertaken for social and political reasons, to fulfill expectations, as well as for reasons of religious belief. Religion, like political and social systems, encompasses a system of both beliefs and behaviors, including values, knowledge and ritual. A common distinguishing mark for religion, as opposed to other systems, is that it is a system of beliefs and behaviors relating to relationships with supernatural beings or forces ("extranatural" forces in Walker's discussion). Archaeologically, it is much easier to reconstruct rituals than beliefs, through the traces left behind by patterned behavior. Our knowledge of ancient religion is thus often skewed in this direction, so we often know about beliefs through analyses of rituals. This is especially the case for societies in which few written or oral accounts are preserved, but even in well-documented cases an archaeological perspective can provide additional insights into behavior patterns and beliefs.

If religious ritual is modeled as a technological system, then the end-product is the desired aim of the ritual, whether rain, healing of disease, or life after death. The objects, actions, and music used in the ritual are various types of tools and materials used to create the end-product through a series of stages during production. These objects, actions, and music require preparation of their own, just as the preparation of materials and tools is part of other sorts of production. A variety of techniques might be used to create the same end-product, or there may only be one process of production that will result in a proper ritual being created. The organization of production includes the way the production sequence is ordered and managed by ritual specialists and all other members of the community involved. Modeling ritual as a technological system is particularly interesting as it is a very clear example of the *process* of the production itself being an essential part of the successful creation of the end-product. There are similar examples of the importance of proper process in the creation of material objects, one of the most well-known being the traditional production of samurai swords.

RELIGIOUS MURAL CONSTRUCTION, USE, AND DISCARD

Solometo (1999; 2000) employs a *chaîne opératoire*-based approach to analyze the creation, use, and disposal of religious mural paintings in their social context. Her data comes from ethnographies and other historic accounts of murals created in religious structures (*kivas*) during rituals by the Pueblo people of the American Southwest (Figure 6.8).

The Pueblo people of Arizona and New Mexico were so named from their practice of living in villages or towns ("pueblo" in Spanish) composed of clusters of multistory stone or adobe room-blocks arranged around a central plaza, a practice still existing to a limited extent today. These communities, particularly the Hopi and Zuni, were intensively studied by early American ethnographers in the late 1800s and early 1900s. First recorded by Spanish explorers in the early 1500s, many communities continue today, and there are well-studied links between the historic Pueblo people and earlier prehistoric traditions defined by archaeologists, such as the Ancestral Pueblo (formerly Anasazi) and Mogollon. Both the historic and prehistoric communities were primarily horticulturalists, growing maize, beans, and squash. The religious beliefs of the historic and modern Pueblo peoples have particularly been a focus of anthropological and archaeological interest, both for their own sakes and as a way to understand religious beliefs prior to Spanish contact. Plog and Solometo (1997) discuss the paradox between the common view of Hopi and Zuni religious beliefs as fairly conservative and enduring, and the evidence for change in society and iconography, especially between the 1300s and 1700s AD/CE.

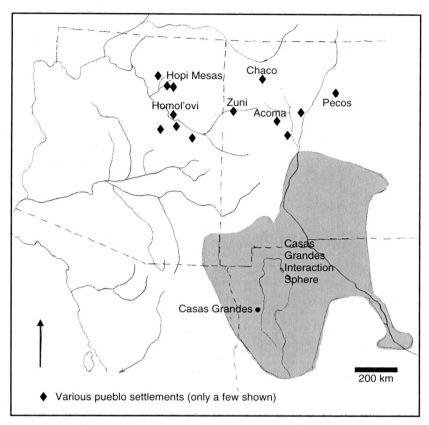

FIGURE 6.8 Map showing Pueblo regions and Casas Grandes Interaction Sphere. (Drawn after maps in Plog and Solometo 1997 and Walker 2001.)

In her research on the murals, however, Solometo is concerned not with change but with extracting a description of the process of mural painting from published ethnohistoric accounts. Her sources were travellers and government agents, ethnographers and archaeologists. Most of the information is from the Hopi and Zuni between the 1880s and 1900, but she also has some data from other pueblos from the mid-1800s to the 1940s. Solometo points out that most past studies of the murals focused on the attributes of the images, primarily to determine their meaning. Instead, Solometo focuses on the production process, to examine the social and religious contexts of the *process* of image-making. This focus on production, she feels, will provide stronger analogies linking ethnographic and archaeological information, by examining the murals as part of a coherent ceremonial complex. Although her study is focused on the murals, she also discusses other ceremonial objects, pointing out that paintings are just one part of an assemblage of ritual items.

Solometo (1999) draws on the work of Lechtman and Lemmonier to create her methodology, particularly the *chaîne opératoire* approach of breaking down the production process into operational stages, as described by Lemmonier (1992). For each stage of production, she discusses (1) the materials and tools used; (2) the actions taken and people involved, noting both gender and relative age; (3) the knowledge needed, both physical and ritual; and (4) associated cultural attitudes or meanings.

The generalized sequence Solometo creates is a summary made up of data from a number of pueblos and ceremonial societies, so it is necessarily a broad outline. The stages of production are as follows:

1. Deciding when to paint
2. Replastering and whitewashing of (wall) surface
3. Painting of designs on kiva rafters
4. Preparation of pigments
5. Preparation of brushes
6. Painting images
7. Destruction or replastering.

Note that stages 2 through 5 are all stages of material preparation—it is not until stage 6 that work on the actual "object," the wall painting, is begun. As seen in Chapters 3 and 4, it is not unusual for material acquisition and preparation to require far more time and energy than the work on the object itself.

The various material preparation stages are by no means similar, in terms of the availability of materials, the people involved, the knowledge needed, and the religious significance. Replastering of the walls involves both men and women in the process, although women typically did the actual replastering itself. It seems to have been a rather lighthearted occasion, although definitely considered to have religious significance, with special materials added to the plaster for various ritual reasons. Preparation of brushes, on the other hand, was not ritually significant; indeed, Solometo (1999: 14) notes that this is the only stage in the entire production process that was not "ritually charged." This is in great contrast to the preparation of the pigments, the most serious and ceremonially-elaborate stage of production other than the actual painting of the murals themselves. Acquiring some of the pigments was extremely difficult, especially the most sacred, and the knowledge of correct preparation methods was held by a few ritual specialists, older men. Less rare pigments might be used by anyone, for everyday decoration as well as ceremonial uses, but for religious applications would likely be mixed with other religiously-valued materials to make the pigment, such as flowers, corn meal or carbonized corn from archaeological sites, and powdered turquoise. If the materials were ground by girls under the direction of the ceremonial specialists, the process took place in a ceremonial atmosphere with appropriate dress, but preparation was often done secretly by the specialists themselves as part of ceremonial rites.

It is thus significant that the painting of mud designs on the kiva rafters is seen as a preparatory stage, done during replastering and separate from the painting of the images on the walls of the kiva. This stage is only described by one 1893 source for the Hopi, but it seems to have been a very regular and faithfully followed tradition, as Solometo (1999) notes that she observed these designs herself at the Hopi village of Walpi in 1997. A different material is used, mud rather than pigment, and different personnel are involved both by gender and age, girls rather than older men or younger men supervised by older men. The painting of mud designs on the rafters also appears to have taken place in a more public setting than the painting of the mural images. The latter were painted as part of ceremonies, but in private preparations prior to any more public parts of the ceremonies.

The mural painting process itself is an essential part of achieving the desired function of the murals (curing, rain-making, etc.); the act of the ritual is likely as important as the final image produced (Solometo 1999, 2000). Solometo describes this stage in great detail, elaborating on the timing, the painters, the viewers, the knowledge involved in image selection and placement, and the meaning and role of the images. She also differentiates between temporary and permanent paintings. Temporary paintings were made and used only for a specific ceremony and then discarded by being covered over or destroyed. Permanent paintings were "left up for a considerable period of time and were periodically renewed or refreshed" (Solometo 1999: 7). She also compares the personnel involved in mural making with those involved in other sorts of dry painting, such as sand and meal paintings (Solometo 1999: 18–19). Like the creation of these Pueblo murals, the creation of sand paintings as part of ritual ceremonies in the traditional religions of the American Southwest as well as in Tibetan Buddhism similarly focuses on the process of production rather than the end-product. This is in contrast to the creation of religious paintings in modern and historic European Christianity, for example, where the final painting was the goal of the production process, and the process itself was generally not considered to be part of a religious ritual. Even stronger examples of the importance of the act of production rather than simply the existence of a final product are seen when dance and music production is considered. In these technologies, there is no final object; the production of dance or music exists only in the process, reinforcing the attention to gesture emphasized in the classic *chaîne opératoire* approach (Lemonnier 1992).

One of the most interesting aspects of the case of Pueblo religious mural production is the importance of proper discard, something that seems to be a common issue in religious ritual processes in many parts of the world, as discussed further below. Solometo points out that proper discard of the paintings is and was very much a part of their production process. In one recorded historic case, paintings were effaced from the walls soon after the ceremony was over. However, at least some prehistoric murals seem to have been primarily "discarded" through re-plastering rather than destruction. Solometo notes that murals discovered by archaeologists typically have many layers of paintings, from dozens to more than 100 layers. At least in the historic case, ethnographers have suggested that discard is so important because the paintings, like other religious objects, contain religious power which must be properly controlled or employed to prevent danger to the community (Solometo 1999: 31-32, 13; 2000).

Archaeological Identification of Religious Ritual

Walker (2001) also emphasizes the importance of proper disposal of ritual objects as a key part of religious ritual for earlier people living in the southern portions of the American Southwest and northern Mexico. He uses the characteristics of disposal technique to identify religious ritual behaviors represented by archaeological deposits. Walker begins with a short but very useful summary and critique of major approaches to religion used by anthropologists, including Tylor, Durkheim, and Eliade. He discusses the implications of each of these approaches for the study of religious ritual and technology, drawing on the work of Horton. In particular, he defines religion as "social relationships or interactions between people and spiritual forces outside of the material world," as I have (Walker 2001: 90). Walker points out that by approaching religion in this way, he can view religious practice as both ideological and pragmatic, and religious artifacts and behaviors can be viewed as part of a technological system. He then provides a brief summary and critique of behavioral archaeology, and applies it to religious ritual technology, employing astute descriptions of the use of analogical reasoning between experimental, ethnographic, and archaeological data to infer the life histories of objects. Walker creates alternative models of object life histories and particularly focuses on disposal techniques, some religious and some not, then evaluates archaeological finds against these models.

Most strikingly, Walker contrasts the possible life history models of pueblo houses impacted by war, domestic accident, and ritual abandonment, showing that the different technological systems involved result in abandonment remains that are different in archaeologically discernable ways. For example, he notes that for a house that was ritually abandoned and burned, one might find whole ritually-important objects placed above the floor, above the collapsed roofing, and in any subsequent garbage fill or wind-blown deposits. Such a depositional sequence is much less likely in a house burnt and abandoned due to warfare. Walker applies such life history models to the case of the social, political, and religious system of the Casas Grandes Interaction Sphere of northern Mexico, southern Arizona, and southwestern New Mexico, for the thirteenth and fourteenth centuries AD/CE (Figure 6.8). He provides a number of examples of archaeological deposits that are most parsimoniously explained as resulting from ritual abandonment, as well as evidence for long-term use of abandoned ceremonial spaces for disposal of religious objects.

The development of alternative life history models for objects seems a particularly fruitful path for the investigation of potential religious objects or traces left by ritual behaviors. At least in the American Southwest for the past thousand years, disposal of religious objects seems to often involve special behaviors. In many modern religions, Christianity included, proper disposal of religious materials is important as a sign of respect, as well as for avoidance of future misuse of the objects and containment of religious power. In contrast, an account of baked clay figurines used in religious rituals in modern Gujarat, India, shows an apparent lack of concern for their subsequent reuse or disposal (Shah 1985). After their use as ritual offerings to spiritual beings to request healing, bestow fertility, or solve some other problem, these figurines are left at the shrine until swept aside by ritual specialists, when they might be left in a heap or picked up for use in play by children. The figurines thus might function both as religious objects and toys for children at different stages in their life history, so alternative models of their use and discard would need to take these possibilities into account.

The complexity of the situations described above should not be taken as a warning that ethnographically described cases are only useful for analyzing archaeological cases from the same group of people in the relatively recent past, and that we cannot investigate religious ritual in other cases. Neither Walker nor Solometo propose that actions or beliefs were necessarily the same in the past as in the present. On the contrary, they both suggest methods for investigating what actually may have occurred in the past, rather than assuming continuity. Plog and Solometo (1997) specifically address a case where they propose that some aspects of the ancient religious ritual system may have been quite different from the ethnohistoric system. They discuss social and ritual change in the Western Pueblos of the American Southwest for the period from the thirteenth through eighteenth centuries, overlapping temporally with Walker's research but in an area farther north. They draw on recent studies of changing iconography and ceremonial architecture by Adams, Crown, and Schaafsma, to analyse the possible role of the emerging katsina religious rituals. Their main goal is to show that the new religious actions and beliefs were in part related to increased conflict occurring at this time, contrary to earlier studies emphasizing the role of these rituals in promotion of rain, crop fertility, and social integration. They do not argue that integration and fertility are unimportant, and they also mention other likely factors relating to change, such as the increase in cotton use which required increased water supply, but their focus is on the role of conflict. Plog and Solometo argue that differences as well as similarities need to be examined to understand the changes as well as the continuities between religious rituals in the prehistoric and recent past. The discussion around this issue continues for the prehistory of this region, but here I want to examine their argument from a different perspective.

Plog and Solometo's (1997) paper predates Solometo's (1999; 2000) research using the *chaîne opératoire* approach to investigate ritual activities as a technological system. Could re-framing the questions asked by Plog and Solometo from a technological perspective provide additional insights into the motivations for religious change? If religious rituals are seen as technologies, the end-product or aim of these technologies is usually stated to be the production of rain, the production of fertile crops, the healing of an ill individual, and so forth. The various operational stages of the ritual can be modeled as for other technologies, as in as the creation of mural paintings described above or similar production sequences for the creation of ritual dances, sand paintings, or other sorts of offerings. The way these production sequences are organized—the personnel involved, the order of stages, the location of stages (public or private, etc.)—result in or are related to social, economic, and political relationships within the community. The particular organization of ritual production can result in the sorts of relationships that are referred to by some Southwestern researchers as "latent functions": economic cooperation to redistribute food; incorrect social behavior publically highlighted through mockery; social and political unification of the community through the need for multiple groups to participate in one or a series of rituals (Plog and Solometo 1997). By modeling this system as a technology, the relationship between the aims of the rituals and their "latent functions" are clearly shown; the aims are the end-product of the rituals, while the "latent functions" are the outcome of the way the ritual production is organized. Thus, as is clear from the discussion in Chapter 5 on Technology and Style, the end-product of the production might remain the same through time or across space, yet the organization of production (the "latent functions" or socio-cultural relationships) might be quite different. Changes in the rituals (technological innovation and adoption) might affect the final aim, but quite often the final aim might be the same—healing, fertility or rainfall—with the new technology perceived as a more effective way of achieving these aims. Whether the aims change or not, the new ritual technologies almost certainly will result in, result from, or relate to differences in the organization of production, so that new social, cultural, economic, or political relationships will occur. This situation highlights Plog and Solometo's point about the importance of change as well as continuity in religious ritual.

I also see some similarity to discussions about technological style in Plog and Solometo's brief discussion of the need to distinguish between "ever-changing" and "never-changing" aspects of Western Pueblo ritual. They themselves cite Rappaport's use of these terms, explaining how their own use is different, but the concept as they use it reminds me of attempts to identify limitations in the degree of choice possible for technological operations. As discussed in Chapter 5, the examination of technological style revolves around techniques or stages of production that can be undertaken in more than one way while providing similar outcomes; the choice of approach made relates to economic, cultural, social, or political factors rather than strictly technological requirements.

This brief example of the application of a technological model to the modeling of religious ritual change will not necessarily provide better insights than other approaches, but it does provide alternative insights and so is worth exploring. For example, there are other cases of depositional patterns that lend themselves to similar analyses, as seen in Russell's (2001b) discussion of the deposition of certain types of still-useful bone points at Çatalhöyük. It is important to include ritual technologies in our analysis of craft production in societies, to fully understand the importance of these crafts, whether from a functionalist, energy-consuming point of view, or to understand the social status of the practitioners of these crafts, or to include the political importance of their products in power systems. This point has been made for years in the ethnographic literature, through numerous comments on the high value and exclusive ownership of the production of ritual items such as songs and dances. As archaeologists have seen in the cases of textiles and gender, the difficulty of recovering information about the perishable and the intangible requires ingenuity, not denial.

Chapters 5 and 6 have illustrated that innovation, maintenance of tradition, style, exchange, ritual, and many other topics central to archaeological understanding of the past can all be effectively studied through the lens of technology. As I will conclude in Chapter 7, the study of multiple technologies allows an even more powerful analysis of the complexities of past societies. Trajectories seen in one craft may not be reflective of the changes occurring in a society as a whole, and this discrepancy can be misleading if only this craft is examined, or insightful if multiple crafts are examined so that this differing trajectory can be recognized and investigated.

CHAPTER 7

The Analysis of Multiple Technologies

In the end, comparison is the key-word.
(Lemonnier 1992: 23)

How can an examination of multiple technologies be useful? After all, it is difficult enough to master the necessary literature on one technology, how can a more superficial knowledge of several be at all useful? One might say the same about the process of archaeology in general—how can one person evaluate and weave together the information from a team of specialists? The answer is that we *must* do so, to have a well-rounded picture of the past. This is the case whether comparing technologies based on material or end-product type, or technologies focused on different processes or functions.

CROSS-CRAFT PERSPECTIVES

Cross-craft comparisons—that is, comparisons between two or more craft technologies—are relatively rare in any part of the world, even in periods with detailed historical records. However, there are some outstanding examples available, and more are being produced all the time. Book-length archaeological examples of cross-craft comparisons include Underhill's (2002) examination of the production and use of food and food containers (pottery and bronze vessels) for ancient China, and Sinopoli's (2003) analysis of pottery, textiles, poetry, and several other crafts for medieval South Asia. Numerous studies are summarized in McGovern and Notis's (1989) edited volume *Cross-Craft and Cross-Cultural Interactions in Ceramics*, including both clay- and silicate-based ceramics. McGovern's introduction to this volume provides an outline of the possible mechanisms for exchanges of style or production techniques that might occur between crafts, and how such exchanges relate to the larger issues of innovation and tradition. Izumi Shimada's forthcoming edited volume (University of Utah Press) on cross-craft production provides examples from a wide variety of crafts and regions. Cross-craft studies produced in other disciplines are also of great value for archaeologists, such as Frank's (1998) ethnographic study which not only describes the production systems of Mande potters and leather-workers, but also contrasts the social status and roles of potters, leather-workers, blacksmiths and other craftspeople among this West African group. And of course, there are the great compendiums of crafts for Mesopotamia by Moorey (1994) and for Egypt edited by Nicholson and Shaw (2000), both inspired by Alfred Lucas' groundbreaking work in Egypt, not to mention the older encyclopedic works by Hodges, Forbes, Singer and company.

In this last chapter, I rather ambitiously want to discuss the creation of a framework for such cross-craft comparisons. For now, I will limit myself to the comparison of different crafts within one society or tradition, and not directly discuss such issues as the adoption or copying of production techniques or organizational methods between groups in the same or different crafts, although that is certainly a type of within-craft or cross-craft comparison that might be very informative. The use of a systematic framework of analysis will, I hope, ultimately allow us to see interactions between different groups of craftspeople, as well as provide information about societies in general.

The study of potential interactions between craftspeople has primarily focused on two aspects of crafts: *style* and *technological style*, as defined in Chapter 5. The vast majority of research on interaction between different crafts has focused on style, particularly on the transmission of designs; for example, the use of designs from textiles or basketry on pottery. The enormous literature on style provides numerous examples of ways that style might reveal interactions between craftspeople. A particular type of design transmission is *skeuomorphism*, the copying of shapes and designs from one material to another. Shapes or surfaces typical of either basketry or metal vessels can be imitated in the production of clay vessels (e.g., Rawson 1989; Vickers 1989), or a surface appearance imitating clay shaping techniques and styles might be incorporated in the production of metal vessels (e.g., Steinberg 1977 for Shang). Another example would be the transfer of styles developed for wooden architecture to stone architecture. Skeuomorphism includes not only imitation of designs, but also imitations of the design of production techniques. For example, wooden architectural production techniques like the joining of wooden beams by lashing with fibers might be imitated by carving stone beams to look as though they were lashed together with fibers. Skeuomorphism does not include the transfer of *actual techniques* used in wood working to stone working. However, there can be aspects of technological innovation to such stylistic interactions. To transfer designs or shapes to a new material, new techniques must frequently be developed or applied, so the adoption of new styles can also result in the development and adoption of new techniques, although not usually the transfer of new techniques from the imitated craft. An excellent example is the copying of silver vessels in porcelain in tenth and eleventh century China. In order to create shapes in clay similar to those created by working sheet metal, the potters did not borrow techniques from metal working, but rather developed new techniques and tools of their own—the exploitation of new clays, the use of new firing techniques and tools, and the use of molds (Rawson 1989). Furthermore, new technologies can also encourage new styles; d'Albis (1985) discusses how the development of new materials, soft and hard porcelain pastes, influenced the development of new ceramic styles in eighteenth century AD/CE Europe.

Information about cross-craft interactions can also be gathered from analyses of technological style. The study of technological exchange between different crafts is much less common than the study of stylistic exchange between crafts, as the study of technological style is proportionally less common than the study of style. There is a good reason for this, as I mentioned on the first page of this volume—it is much more difficult to do, especially from archaeological remains alone. To become expert enough in one craft to make sense of archaeological remains is a time-consuming challenge; to know so much about more than one craft is that much more demanding. The types of tools, techniques and organizational methods used by a single group of craftspeople are complex and elusive enough without trying to compare such aspects of different crafts. And from a practical perspective, there are fewer chances of having the available archaeological contexts to study two or more crafts. This is why I advocate team approaches wherever possible, although it is important to have team members try to learn about each other's crafts, not just each other's conclusions.

TECHNOLOGICAL STYLE AND CROSS-CRAFT INTERACTIONS

So what sorts of clues to cross-craft interactions can we find in *technological styles*? I divide my discussion into three groups, following my nested definition of technology: *tools or materials, techniques,* and *organizational methods.* The examples I use here are ways of looking for the points in the craft production processes which seem to offer the most promise for future research on potential interaction; in other words, finding an end of the string as the first step to unraveling this tangle of cross-craft comparisons, not the full unraveling, much less the organized skein of yarn. I want to stress that it is necessary to look at *both* differences and similarities between the tools, techniques, and organizational methods of crafts to get at both possible interactions between craftspeople and information about their societies. I repeat again van der Leeuw's (1993) stricture that we need to look at both the choices made and the choices not made.

A common form of interaction between crafts is the production of specialized *tools* used by one craft for another craft, such as production of clay loom weights by potters for weavers. First, it cannot be assumed that such simple objects as clay loom weights were not made by weavers themselves—such objects have to be carefully studied and compared with known products of the potter. Certainly in societies where large numbers of people are using loom weights, and particularly where weavers are specialist workers, it would be likely that potters might make batches of loom weights. In smaller-scale societies with more self-sufficient households, the weaver might also be the potter or at least have access to the necessary materials (clay) and tools (firing structures) to create the loom weights personally. A second, less common form of interaction between crafts related to tools is the adoption of a tool type used in one craft by a different craft. In general, such borrowing is very difficult to identify, particularly where few or no written records exist. The easiest adoption to identify, adoption of a specialized tool, does not occur very often, as a specialized tool developed for one craft is seldom exactly right for another craft. The use of similar but very simple or generalized tools for similar tasks would be difficult to establish as "adopted" tools—for example, the use of stone polishers by metal, wood, and stone workers. However, the use of an identical type of stone from a restricted source to make the polishers used by all of these industries, especially when other useable stones existed, provides information about the dependence of all of these crafts on one source of supply, and thus a clue to the structure of the economy. The high-temperature pyrotechnological crafts (metal melting and smelting, pottery firing, vitreous material production) all share a common need to create tools and techniques for dealing with the production and control of high-temperature fires. Quite often, these crafts are firing objects at a temperature above the vitrification point of the clays used to make the firing structures and associated tools (crucibles, setters, containers). A clue to the degree of interaction between craftspeople is thus whether or not they used the same methods to solve this problem, all else being equal in the production process. Therefore, the development of refractory materials is often a place where common tools are sought by archaeologists working on these crafts, and a number of publications exist on this topic, such as the range of articles on refractories published in McGovern and Notis (1989). However, all else is not always equal in these production processes, as different crafts have different atmospheric or handling requirements which may affect the choice of refractory materials, so these difference need to be assessed when deciding if borrowing along with modification of tools is a possibility.

The comparison of *techniques* between crafts is again that much more difficult and requires that much more detailed information about craft production techniques. However, I have already presented a number of examples of the use of the same production technique in three different crafts: the use of groove-and-snap techniques in stone, bone, and North American metal working. The groove-and-snap technique is very simple, is found worldwide, and dates to very early times, the Upper Paleolithic at least for bone and antler work. It is the most efficient technique to use when cutting with stone tools of most types, and so it may fall into the category of a generalized technique, like the generalized tools above, making the borrowing of this technique between crafts and between societies no more (or less) likely than independent invention. However, the continued use of such techniques when new tools usually associated with new techniques are available—the choice *not* to change techniques—is a significant indicator of technological style and the maintenance of tradition. This maintenance of tradition may occur for economic reasons (the tools are not widely available or are expensive), or because learning the new technique is not as efficient as continuing in the old, as Wake (1999) suggested for Native Alaskan sea mammal hunting. When maintenance of a technique is found across several crafts and situations, however, it implies that continued use of this technique may have some particular social or ideational association. Social or ideational reasons should not be given automatic precedence over functional or economic reasons, or vice versa, of course; all have to be equally considered as alternatives.

Another example of the borrowing of a technique between crafts is the use of molds to make clay figurines and faience figurines in ancient Eurasia. Studies of exact molding techniques and materials, as well as archaeological evidence for the appearance of these techniques, can help to determine if this is a borrowed technique. The examination of the use of molds in another craft, metal casting, might provide additional insights. For example, what is it about the use of molds in these crafts that makes it likely that they represent borrowing of a technique from clay working to faience working, but not the borrowing of the same technique from clay working to metal casting? We also see many societies in which metal is cast, but molding is never used as a method of production for clay objects. This *lack* of borrowing of a technique is also important to investigate, in order to understand the perceptions about these techniques and degrees of interaction between crafts within the given society, and in order to understand cases where borrowing does occur. Of course, borrowing of techniques is not the only way crafts can interact with and influence each other. They can actually set up systems of co-dependence in production, such as textile workers having their cloth processed by fullers or dyers.

Finally, like many of my colleagues, I am interested in trying to compare the *organizational methods* of one craft with another (Vidale and Miller 2000). Such comparisons form the root of many archaeological attempts to examine social or political control of one craft versus another. One method which has been used for a long time to examine the process and organization of production for various crafts, particularly lithics, is to sketch out the stages of production in a standardized framework, such as I have used for each of the sections in Chapters 3 and 4 (Figures 3.1, 3.13, 3.20, 4.3, 4.8, 4.11). The best known use of this technique is the *chaîne opératoire* approach, as discussed in Chapter 2, but very similar frameworks are used in models derived from behavioral archaeology (Skibo and Schiffer 2001), operations process management (Bleed 1991), and information systems analysis (Kingery 1993). The frameworks provided in Chapters 3 and 4 are highly simplified overviews for an entire craft process; much more specific pathways can be produced for the production of one specific type of stone object to compare it with another. However, diagramming the entire craft process is a good way to step back from the details and see the overall process of metal production, for example, in order, to compare it with the overall process of pottery production (Figures 7.1 and 7.2).

Using frameworks like this, it is easy to see that there are no semi-finished products produced during pottery production, after the preparation of the clay body. That is, there are no easily movable, storable, semi-finished goods, which can be shipped to another location for working, or stockpiled and altered to produce objects needed at a later date. Instead, all of the stages of pottery production after materials preparation need to take place in a fairly spatially restricted area and preferably within a relatively short time period, as the unfired products are extremely fragile. In contrast, copper production offers several stages at which semi-finished products are produced: smelting ingots, refined or alloyed ingots, cast blanks (bars or disks of various kinds), and even scrap. Semi-finished products can be shipped long distances, allowing for production at centers far distant from the sources of raw materials. This search for potential semi-finished products can thus be used as a method of predicting potential points of segregation within craft production processes. Such an easy separation of various stages of production allows for or encourages specialization of craftspeople in particular stages, so that one person no longer produces an object from start to finish. Indeed, most craftspeople probably did not have the skill and/or knowledge for some of the most elaborate craft production sequences, such as the manufacture of embroidered brocades. While more likely in large-scale societies, segregation can also easily occur

The Analysis of Multiple Technologies 243

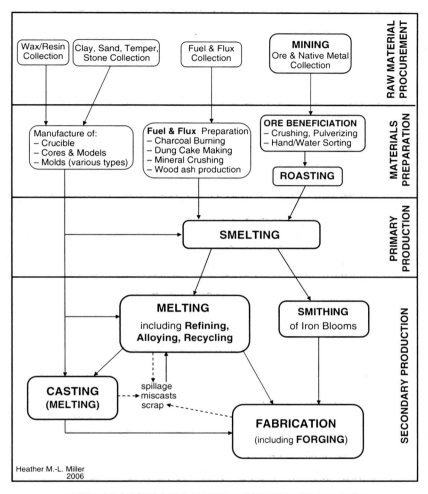

PRODUCTION PROCESS DIAGRAM FOR COPPER AND IRON

FIGURE 7.1 Generalized production process diagram for copper and iron (greatly simplified).

in very small-scale societies; as MacKenzie (1991) has documented, the processed plant fibers (*tulip*) shown in Figure 3.14 were traded several days walk through the mountains of Papua New Guinea for use in string bag (*bilum*) manufacture by women in other groups. Crafts without semi-finished products also had spatially segregated stages carried out by different specialists, of course, such as the systems of organization for large-scale pottery production discussed in the Labor section of Chapter 5. Even in these cases, however,

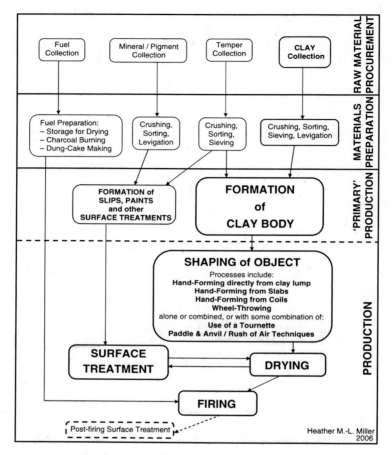

PRODUCTION PROCESS DIAGRAM FOR FIRED CLAY (pottery)

FIGURE 7.2 Generalized production process diagram for fired clay, focused on pottery (greatly simplified).

production process diagrams help us to see where such segmentations might occur, such as between shaping and surface treatments for pottery production.

This separation of craft production stages when combined with larger-scale production can in turn encourage the growth of a managerial group, who smooth the flow of production and coordinate efforts, procure necessary raw and semi-finished materials, and maintain quality levels. In small workshops, this would probably be done by the master/owner, who was likely a skilled worker. In the largest workshops, managers may not be skilled workers at all, but would necessarily be individuals who understood the process and could judge the quality of both materials and products. Two such systems were studied for stone bead production by Kenoyer, Vidale, and Bahn, as discussed in Chapter 3. The temporal aspects of semi-finished object production would encourage the development of a manager with particular knowledge of the demand for items both locally and at a distance. This is the case for large merchant houses, but also for individual producers trading in small-scale societies, even nomadic herders carrying minerals between the ends of their transhumance pattern. Semi-finished objects can be stockpiled, then the desired type of object produced when demand is most favorable to the producer or trader. One would thus expect to see quite different methods of organizing, and by extension controlling, crafts with potential semi-finished products and those without. This hypothesis can be extended to other crafts, to see if we can generally say that crafts with potential semi-finished products (such as metals, cloth, glass, and chert blade production) are organized differently from crafts without this option (such as pottery and faience production). Note that this is not a simple difference between crafts with relatively complex and crafts with relatively simple production processes.

This is not a model—this is a starting place. Many of the points suggested by these methods seem obvious for well-studied or often-compared crafts like copper object and pottery production, but are extremely useful for the tangle of vitreous materials, or for the social information that might be available in comparing apparently quite different crafts. Cross-craft comparisons can be extremely helpful in assessing relative value, as discussed in Chapter 6, particularly if technological processes may be part of this value. Cross-craft comparisons can also be made across regions, across scales of societies, and across types of social and political organization. After all, the reality of the world, present and past, is of a mosaic of groups interacting with each other, hunter-gatherer groups trading forest products to urban dwellers for pottery or metal, or small fishing villages participating in multistate exchange networks, so that craft products and perhaps techniques cross geographical and social boundaries every day. Cross-craft technology studies allow a great deal of freedom in comparing groups, and the way they create, adopt, and employ technologies.

This introduction to the way archaeologists approach technology has, I hope, piqued interest in this fascinating, interconnected, endless field of study. May good journeying through this country be yours, whether you are following paths or blazing them.

> ... the past has a way of luring curious travelers off the beaten track. It is, after all, a country conducive to wandering, with plenty of unmarked roads, unexpected vistas, and unforeseen occurrences. Informative discoveries, pleasurable and otherwise, are not at all uncommon.
>
> (Basso 1996: 3–4)

BIBLIOGRAPHY

Adams, Jenny L.
2002 *Ground Stone Analysis. A Technological Approach*. Salt Lake City: University of Utah Press, in conjunction with the Center for Desert Archaeology, Tucson.

Agricola, Georgius
1950 [1556] *De Re Metallica*. Translated from the first Latin edition of 1556. Hoover, Herbert Clark and Lou Henry Hoover. New York: Dover Publications, Inc.

Anderson-Gerfaud, P., M.-L. Inizan, M. Lechevallier, J. Pelegrin, and M. Pernot
1989 Des lames de silex dans un atelier de potier harappéen. Interaction de domaines techniques. *Comptes Rendus de l'Académie des Sciences* series 2, 308:443–449.

Andrefsky, W.
1998 *Lithics: Macroscopic Approaches to Analysis*. Cambridge: Cambridge University Press.

Anselmi, Lisa Marie
2004 *New Materials, Old Ideas: Native Use of European-Introduced Metals in the Northeast*. PhD thesis, University of Toronto. Available through UMI Publications, Digital Dissertations.

Archaeology Branch
2001 *Culturally Modified Trees of British Columbia; A Handbook of the Identification and Recording of Culturally Modified Trees*. Victoria, BC, Canada: Ministry of Small Business, Tourism and Culture, Province of British Columbia.

Arnold, D. E.
1985 *Ceramic Theory and Cultural Process*. Cambridge: Cambridge University Press.

Arnold, Jeanne E.
1995 Transportation Innovation and Social Complexity among Maritime Hunter-Gatherer Societies. *American Anthropologist* 97(4):733–747.

Arnold, Jeanne E.
2001 The Chumash in World and Regional Perspectives. In: *The Origins of a Pacific Coast Chiefdom. The Chumash of the Channel Islands.* Arnold, Jeanne E., ed. Salt Lake City: The University of Utah Press. pp. 1–19.

Bachmann, Hans-Gert
1982 *The Identification of Slags from Archaeological Sites.* Institute of Archaeology Occasional Publication No. 6. London: Institute of Archaeology.

Banning, Edward B.
2000 *The Archaeologist's Laboratory. The Analysis of Archaeological Data.* Interdisciplinary Contributions to Archaeology. New York: Kluwer Academic/Plenum Publishers.

Banning, Edward B.
2002 *Archaeological Survey.* Manuals in Archaeological Method, Theory, and Technique. New York: Kluwer Academic/Plenum Publishers.

Barber, Elizabeth W.
1991 *Prehistoric Textiles.* Princeton, NJ: Princeton University Press.

Barber, Elizabeth W.
1994 *Women's Work: The First 20,000 years. Women, Cloth, and Society in Early Times.* New York: W. W. Norton & Company, Inc.

Barber, Russell J.
1994 *Doing Historical Archaeology. Exercises Using Documentary, Oral, and Material Evidence.* Englewood Cliffs, NJ: Prentice Hall.

Barthélémy de Saizieu, Blanche
2000 Les perles en roches dures du site de Nausharo (Baluchitan pakistanais), 2800–2000 av. J.-C. (Stone beads from the Nausharo site (Pakistani Baluchistan), 2800–2000 B.C.). In: *Cornaline de l'Inde. Des pratiques techniques de Cambay aux techno-systèmes de l'Indus.* Roux, Valentine, ed. Paris: Éditions de la Maison des sciences de l'homme. pp. 333–410.

Barthélémy de Saizieu, Blanche
2004 *Les Parures de Mehrgarh. Perles et pendentifs du néolithique précéramique à la période pré-Indus. Fouilles 1974–1985.* Paris: Éditions Recherche sur les Civilisations.

Barthélémy de Saizieu, Blanche, and Anne Bouquillon
1994 Steatite working at Mehrgarh during the Neolithic and Chalcolithic periods: quantitative distribution, characterization of material and manufacturing processes. In: *South Asian Archaeology 1993.* Parpola, Asko and Petteri Koskikallio, eds. Annales Academiae Scientiarum Fennicae, Series B, Volume 271, Vol. 1. Helsinki: Suomalainen Tiedeakatemia. pp. 47–59.

Barthélémy de Saizieu, Blanche, and Anne Bouquillon
1997 Evolution of Glazed Materials from the Chalcolithic to the Indus Period Based on the Data of Mehrgarh and Nausharo. In: *South Asian Archaeology 1995.*

Allchin, Raymond and Bridget Allchin, eds, Vol. 1. New Delhi: Oxford & IBH Publishing Co. Pvt. Ltd. pp. 63–76.

Basalla, George
1988 *The Evolution of Technology*. Cambridge History of Science Series. Cambridge: Cambridge University Press.

Bass, George Fletcher
1975 *Archaeology Beneath the Sea*. New York: Walker.

Basso, Keith H.
1996 *Wisdom Sits in Places*. Albuquerque, NM: University of New Mexico Press.

Bayley, Justine
1985 What's what in ancient technology: an introduction to high-temperature processes. In: *The archaeologist and the laboratory*. Phillips, Patricia, ed. Council for British Archaeology Research Report 58. London: Council for British Archaeology. pp. 41–44.

Bayley, Justine
1989 Non-Metallic Evidence for Metalworking. In: *Archaeometry. Proceedings of the 25th International Symposium*. Maniatis, Y., ed. Amsterdam: Elsevier Science Publications. pp. 291–303.

Bealer, Alex W.
1976 *The Art of Blacksmithing*. Revised edition. New York: Funk and Wagnalls.

Belcher, William Raymond
1994 Multiple Approaches towards Reconstruction of Fishing Technology: Net Making and the Indus Valley Tradition. In: *From Sumer to Meluhha: Contributions to the Archaeology of South and West Asia in Memory of George F. Dales, Jr.* Kenoyer, Jonathan Mark, ed. Wisconsin Archaeological Reports, Volume 3. Madison, WI: Department of Anthropology, University of Wisconsin–Madison. pp. 129–141.

Bellina, Berenice
2003 Beads, social change and interaction between India and South-east Asia. *Antiquity* 77(296):285–297.

Biagi, Paulo, and M. Cremaschi
1991 The Harappan flint quarries of the Rohri Hills. *Antiquity* 65:97–102.

Binford, Lewis, and Sally Binford
1966 A preliminary analysis of functional variability in the Mousterian of Levallois facies. *American Anthropologist* 68:238–295.

Biskowski, Martin
1997 *The adaptive origins of prehispanic markets in central Mexico: The role of maize-grinding tools and related staple products in early state economics*. PhD thesis, University of California, Los Angeles.

Bisson, Michael S., S. Terry Childs, Philip de Barros, and Augustin F. C. Holl, eds.
2000 *Ancient African Metallurgy*. Walnut Creek, CA: Altamira Press.

Blackman, M. James, and Massimo Vidale
1992 The Production and Distribution of Stoneware Bangles at Mohenjo-Daro and Harappa as Monitored by Chemical Characterization Studies. In: *South Asian Archaeology 1989*. Jarrige, Catherine, ed. Monographs in World Archaeology No. 14. Madison, WI: Prehistory Press. pp. 37–44.

Bleed, Peter
1991 Operations Research and Archaeology. *American Antiquity* 56(1):19–35.

Bleed, Peter
2001 Artifice Constrained: What Determines Technological Choice? In: *Anthropological Perspectives on Technology*. Schiffer, Michael B., ed. Albuquerque: University of New Mexico Press. pp. 151–162.

Bordes, François
1961 Mousterian cultures in France. *Science* 134:803–810.

Bordes, François
1973 On the chronology and contemporaneity of different Paleolithic cultures in France. In: *The explanation of culture change*. Renfrew, Colin, ed. London: Duckworth. pp. 217–226.

Bouquillon, Anne, and Blanche Barthélémy de Saizieu
1995 Découverte d'un nouveau matériau dans les parures de la période pré-Indus de Mehrgarh (Balochistan): la "faïence" de stéatite. *Techne: La science au service de l'histoire de l'art et des civilisations. Autoportrait d'un laboratoire, le Laboratoire de recherche des musées de France* 2:50–55.

Bowkett, Laurence, Stephen Hill, Diana Wardle, and K.A. Wardle
2001 *Classical archaeology in the field: Approaches*. Bristol: Bristol Classical Press.

Bradley, Bruce
1989 *Flintknapping with Bruce Bradley*. Cortez, Colorado: INTERpark. Videotape, ca. 55 minutes.

Brannt, William T.
1919 *Metal Worker's Handy-Book of Receipts and Processes. Being a collection of chemical formulas and practical manipulations for the working of all the metals and alloys; including the decoration and beautifying of articles manufactured therefrom, as well as their preservation*. New York: Henry Carey Baird and Co., Inc.

Bril, Blandine, Valentine Roux, and Gilles Dietrich
2000 Habiletés impliquées dans la taille des perles en calcédoine: caractéristiques motrices et cognitives d'une action située complexe. (Skills involved in knapping of chalcedony beads: Motor and cognitive characteristics of a complex situated action.) In: *Cornaline de l'Inde. Des pratiques techniques de Cambay aux technosystèmes de l'Indus*. Roux, Valentine, ed. Paris: Éditions de la Maison des sciences de l'homme. pp. 211–329.

Bronson, Bennet
1986 The making and selling of Wootz, a crucible steel of India. *Archaeomaterials* 1:13–51.

Bronson, Bennet
1999 The Transition to Iron in Ancient China. In: *The Archaeometallurgy of the Asian Old World*. Pigott, Vincent C., ed. MASCA Research Papers in Science and Archaeology 16; University of Pennsylvania Museum Monograph 89. Philadelphia: The University Museum, University of Pennsylvania. pp. 177–198.

Brown, Rachel
1987 *The Weaving, Spinning, and Dyeing Book*. Second edition, revised and expanded. New York, NY: Alfred A. Knopf.

Brumfiel, Elizabeth M.
1991 Weaving and cooking: women's production in Aztec Mexico. In: *Engendering Archaeology: Women and Prehistory*. Gero, Joan M. and Margaret W. Conkey, eds. Social Archaeology. Oxford: Basil Blackwell Ltd. pp. 224–251.

Carr, Christopher, ed.
1985 *For Concordance in Archaeological Analysis: Bridging Data Structure, Quantitative Technique, and Theory*. Kansas City, MO: Westport Publishers, Inc.

Carter, Howard, and A.C. Mace
1923 *The Tomb of Tut-Ankh-Amen*. London: Cassell and Company, Ltd.

Chakrabarti, Dilip K., and Nayanjot Lahiri
1996 *Copper and its Alloys in Ancient India*. New Delhi: Munshiram Manoharlal Publishers Pvt. Ltd.

Charles, Michael
1998 Fodder from Dung: the Recognition and Interpretation of Dung-Derived Plant Material from Archaeological Sites. *Environmental Archaeology* 1:111–122. Thematic Issue. Fodder: Archaeological, Historical and Ethnographic Studies. Edited by Michael Charles, Paul Halstead, and Glynis Jones.

Chazan, Michael
1997 Redefining Levallois. *Journal of Human Evolution* 33:719–735.

Childe, V. Gordon
1981 [1956] *Man Makes Himself*. First illustrated edition, reprint of third edition of text. Bradford-on-Avon, UK: Moonraker Press and Pitman Publishing Ltd.

Childs, S. Terry, ed.
2004   *Our Collective Responsibility: The Ethics and Practice of Archaeological Collections Stewardship.* Washington, DC: Society for American Archaeology.

Childs, S. Terry, and David Killick
1993   Indigenous African Metallurgy: Nature and Culture. *Annual Review of Anthropology* 22:317–337.

Chilton, Elizabeth S., ed.
1999   *Material Meanings. Critical approaches to the interpretation of material culture.* Foundations of Archaeological Inquiry. Salt Lake City: University of Utah Press.

Choyke, Alice M., and László Bartosiewicz, eds.
2001   *Crafting Bone: Skeletal Technologies through Space and Time. Proceedings of the 2nd meeting of the (ICAZ) Worked Bone Research Group, Budapest, 31 August–5 September 1999.* BAR International Series 937. Oxford: Archaeopress.

Clark, Anthony
1990   *Seeing Beneath the Soil: Prospecting methods in archaeology.* London: B. T. Batsford Ltd.

Clark, John E., and Stephen D. Houston
1998   Craft Specialization, Gender, and Personhood among the Post-conquest Maya of Yucatan, Mexico. In: *Craft and Social Identity.* Costin, Cathy L. and Rita P. Wright, eds. Archaeological Papers of the American Anthropological Association (AP3A), Number 8. Arlington, VA: American Anthropological Association. pp. 31–46.

Clark, Terence
n.d.   *The Northwest Coast Cedar Bent-wood Box: Traditional Production Sequence and Replication.* Manuscript and Presentation, Ancient Technology graduate seminar, Anthropology Department, University of Toronto.

Cleuziou, Serge, and Maurizio Tosi
1994   Black Boats of Magan: some thoughts on Bronze Age water transport in Oman and beyond from the impressed bitumen slabs of Ra's al-Junayz. In: *South Asian Archaeology 1993.* Parpola, Asko and Petteri Koskikallio, eds. Annales Academiae Scientiarum Fennicae, Series B, Volume 271, Vol. 2. Helsinki: Suomalainen Tiedeakatemia. pp. 745–761.

Collon, Dominique
1990   *Near Eastern Seals.* British Museum Interpreting the Past. 2. Berkeley & Los Angeles: University of California Press.

Conkey, Margaret W., and Joan M. Gero
1991   Tensions, Pluralities, and Engendering Archaeology: An Introduction to Women and Prehistory. In: *Engendering Archaeology: Women and Prehistory.* Gero, Joan M. and Margaret W. Conkey, eds. Oxford: Basil Blackwell. pp. 3–30.

Cooke, Strathmore R. B., and Bruce V. Nielsen
1978 Slags and Other Metallurgical Products. In: *Excavations at Nichoria in Southwest Greece. Volume I: Site, Environs, & Techniques.* Rapp, George and S. E. Aschenbrenner, eds. Minneapolis: University of Minnesota Press. pp. 182–224.

Costin, Cathy L.
1991 Craft Specialization: Issues in Defining, Documenting, and Explaining the Organization of Production. In: *Archaeological Method and Theory, Vol. 3.* Schiffer, Michael B., ed. Tucson: University of Arizona Press. pp. 1–56.

Costin, Cathy L.
1996 Exploring the Relationship Between Gender and Craft in Complex Societies: Methodological and Theoretical Issues of Gender Attribution. In: *Gender and Archaeology.* Wright, Rita P., ed. Philadelphia, PA: University of Pennsylvania Press. pp. 111–140.

Costin, Cathy L.
1998a Housewives, chosen women, skilled men: cloth production and social identity in the Late Prehispanic Andes. In: *Craft and Social Identity.* Costin, Cathy L. and Rita P. Wright, eds. Archaeological Papers of the American Anthropological Association (AP3A), Number 8. Arlington, VA: American Anthropological Association. pp. 123–144.

Costin, Cathy L.
1998b Introduction: Craft and Social Identity. In: *Craft and Social Identity.* Costin, Cathy L. and Rita P. Wright, eds. Archaeological Papers of the American Anthropological Association (AP3A), Number 8. Arlington, VA: American Anthropological Association. pp. 3–18.

Costin, Cathy L.
2001 Craft Production Systems. In: *Archaeology at the Millennium: A Sourcebook.* Feinman, Gary M. and T. Douglas Price, eds. New York: Kluwer/Plenum. pp. 273–327.

Costin, Cathy L., and Rita P. Wright, eds.
1998 *Craft and Social Identity.* Archaeological Papers of the American Anthropological Association (AP3A), Number 8. Arlington, VA: American Anthropological Association.

Cotterell, Brian, and Johan Kamminga
1990 *Mechanics of Pre-industrial Technology.* Cambridge: Cambridge University Press.

Cowgill, George L.
2003 Teotihuacan: Cosmic Glories and Mundane Needs. In: *The Social Construction of Ancient Cities.* Smith, Monica L., ed. Washington D.C.: Smithsonian Books. pp. 37–55.

Craddock, Paul T.
1989 The Scientific Investigation of Early Mining and Metallurgy. In: *Scientific Analysis in Archaeology and its Interpretation.* Henderson, Julian, ed. Oxford University

Committee for Archaeology Monograph 19, UCLA Institute of Archaeology Archaeological Research Tools 5. Oxford: Oxford University Press. pp. 178–212.

Craddock, Paul T.
1995 *Early Metal Mining and Production.* Washington, D.C.: Smithsonian Institution Press.

Craig, Barry
1988 *Art and Decoration of Central New Guinea.* Shire Ethnography, Number 5. Aylesbury, Buckshire, UK: Shire Publications Ltd.

Crawford, Harriet
2001 The British Archaeological Expedition to Kuwait. *Newsletter of the British School of Archaeology in Iraq* 7:10–11.

Cunningham, Jerimy J.
2003 Transcending the "Obnoxious Spectator": a case for processual pluralism in ethnoarchaeology. *Journal of Anthropological Archaeology* 22:389–410.

d'Albis, A.
1985 The History of Innovation in European Porcelain Manufacture and the Evolution of Style: Are They Related? In: *Technology and Style.* Kingery, W. David, ed. Ceramics and Civilization, Volume II. Columbus, OH: The American Ceramic Society. pp. 397–412.

Dales, George F., and Jonathan Mark Kenoyer
1986 *Excavations at Mohenjo Daro, Pakistan: The Pottery.* University Museum Monograph 53. Philadelphia: University Museum.

David, Nicholas, and Carol Kramer
2001 *Ethnoarchaeology in Action.* Cambridge World Archaeology. Cambridge: Cambridge University Press.

Des Lauriers, Matthew R.
2005 The Watercraft of Isla Cedros, Baja California: Variability and capabilities of indigenous seafaring technology along the Pacific Coast of North America. *American Antiquity* 70(2):342–360.

Devonshire, Amanda, and Barbara Wood, eds.
1996 *Women in Industry and Technology from Prehistory to the Present Day: Current research and the museum experience.* Proceedings from the 1994 WHAM conference. London: Museum of London.

Dibble, Harold L.
1995 Middle Paleolithic scraper reduction: Background, clarification, and review of evidence to date. *Journal of Archaeological Method and Theory* 2(4):299–368.

Dobres, Marcia-Anne
2000 *Technology and Social Agency: Outlining an Anthropological Framework for Archaeology.* Oxford: Blackwell.

Dobres, Marcia-Anne, and Christopher R. Hoffman
1995 Social Agency and the Dynamics of Prehistoric Technology. *Journal of Archaeological Method and Theory* 1(3):211–258.

Dobres, Marcia-Anne, and Christopher R. Hoffman, eds.
1999 *The Social Dynamics of Technology. Practice, Politics, and World Views.* Washington DC: Smithsonian Institution Press.

Donnan, Christopher B.
2004 *Moche Portraits from Ancient Peru.* Austin: University of Texas Press.

Doyle, Sir Arthur Conan
1988 "The Boscombe Valley Mystery" In *The Adventures of Sherlock Holmes.* Reprinted in: *The Complete Sherlock Homes.* With a Preface by Christopher Morley. Garden City, NY: Doubleday & Company, Inc. pp. 202–217.

Dransart, Penny
2002 *Earth, Water, Fleece, and Fabric: An Ethnography and Archaeology of Andean Camelid Herding.* London: Routledge.

Drennan, Robert D.
1996 *Statistics for Archaeologists: A commonsense approach.* New York: Plenum Press.

Drewett, Peter
1999 *Field Archaeology: An introduction.* London: UCL Press.

Drooker, Penelope Ballard, and Laurie D. Webster, eds.
2000 *Beyond Cloth and Cordage: Current Approaches to Archaeological Textile Research in the Americas.* Salt Lake City: University of Utah Press.

du Bois, Ron
1972 *The Working Process of the Potters of India: Bindapur – A colony of 700 potters*: Oklahoma State University.Videotape, 30 minutes.

Durkheim, Émile
1933 [1893] *The Division of Labor in Society.* Translated by George Simpson. New York, NY: Macmillian.

Ehrhardt, Kathleen L.
2005 *European Metals in Native Hands. Rethinking Technological Change 1640–1683.* Tuscaloosa, AL: University of Alabama Press.

Emery, I.
1980 *The Primary Structures of Fabrics; an illustrated classification.* Second edition. Washington DC: The Textile Museum.

Emmerich, A.
1965 *Sweat of the Sun and Tears of the Moon: Gold and Silver in Pre-Columbian Art.* Seattle: University of Washington Press.

Ericson, J. E., and Barbara A. Purdy, eds.
1984 *Prehistoric Quarries and Lithic Production.* Cambridge: Cambridge University Press.

Fannin, Allen
1970 *Handspinning. Art and Technique.* New York, NY: Van Nostrand Reinhold Company.

Feinman, Gary
1986 The Emergence of Specialized Ceramic Production in Formative Oaxaca. In: *Economic Aspects of Prehispanic Highland Mexico.* Isaac, Barry L., ed. Research in Economic Anthropology, Supplement 2. Greenwich: JAI Press. pp. 347–373.

Finney, Ben
1998 Traditional Navigation and Nautical Cartography in Oceania. In: *The History of Cartography, Vol. 3, part 2: Cartography in Traditional African, American, Arctic, Australian, and Pacific Societies.* Lewis, G. Malcolm and David Woodward, eds. Chicago: University of Chicago Press. pp. 443–492.

Finney, Ben
2003 *Sailing in the Wake of the Ancestors. Reviving Polynesian Voyaging.* Honolulu, HI: Bishop Museum Press.

Fleming, Stuart
1999 *Roman Glass. Reflections on cultural change.* University of Pennsylvania Museum of Archaeology and Anthropology. Philadelphia, PA: University Museum Publications.

Forbes, Robert James
1964–1972 *Studies in Ancient Technology.* Second revised edition. 9 volumes. Leiden: Brill.

Foreman, Richard
1978 Disc Beads: Production by Primitive Techniques. *The Bead Journal* 3(3,4):17–22.

Fox, Christine
1988 *Asante Brass Casting: Lost-wax casting of gold-weights, ritual vessels and sculptures, with handmade equipment.* Cambridge African Monographs 11. Cambridge: Cambridge University African Studies Center.

Francfort, H.-P.
1989 *Fouilles de shortugaï recherches sur l'Asie centrale protohistorique.* Paris: Diffusion de Boccard.

Frank, Barbara E.
1998 *Mande Potters and Leather-workers. Art and Heritage in West Africa.* Washington D.C.: Smithsonian Institution Press.

Frank, Susan
1982 *Glass and Archaeology.* London, New York: Academic Press.

Franklin, Ursula
1992   *The Real World of Technology*. Canadian Broadcasting Corporation (CBC) Massey Lecture Series. Originally published in 1990 by CBC Enterprises. Concord, ON: House of Anansi Press Ltd.

Franklin, Ursula M., E. Badone, R. Gotthardt, and B. Yorga
1981   *An Examination of Prehistoric Copper Technology and Copper Sources in Western Arctic and Subarctic North America*. National Museum of Man Mercury Series Paper No. 101. Ottawa: National Museums of Canada.

Freestone, Ian C.
1989   Refractory Materials and their procurement. In: *Old World Archaeometallurgy*. Hauptmann, Andreas, E. Pernicka, and G.A. Wagner, eds. Bochum, Germany: Dr Anschnitt. pp. 155–163.

Frink, Lisa, and Kathryn Weedman, eds.
2005   *Gender and Hide Production*. Gender and Archaeology Series, Vol. 11. Walnut Creek, CA: Altamira Press.

Fröhlich, Max
1979   *Gelbgiesser im Kameruner Grasland*. Zürich: Rietberg Museum.

Galaty, Michael L.
1999   *Nestor's Wine Cups: Investigating Ceramic Manufacture and Exchange in a Late Bronze Age "Mycenaean" State*. British Archaeological Reports International Series No. 766. Oxford: British Archaeological Reports.

Gamble, Lynn H.
2002   Archaeological Evidence for the Origin of the Plank Canoe in North America. *American Antiquity* 67(2):301–315.

Gero, Joan M., and Margaret W. Conkey, eds.
1991   *Engendering Archaeology. Women and Prehistory*. Oxford: Basil Blackwell.

Gies, Frances and Joseph
1994   *Cathedral, Forge, and Watherwheel. Technology and Invention in the Middle Ages*. New York, NY: HarperCollins Publishers Inc.

Glassie, Henry
1999   *Material Culture*. Bloomington, IN: Indiana University Press.

Goffer, Zvi
1996   *Elsevier's Dictionary of Archaeological Materials and Archaeometry: In English with translations of terms in German, Spanish, French, Italian, and Portuguese*. Amsterdam, New York: Elsevier.

Goldstein, David J. and Izumi Shimada
in press   Middle Sicán Multi-Craft Production: Resource Management and Labor Organization. In *Rethinking Craft Production: The Nature of Producers and Multi-Craft Interaction*. Shimada, Izumi, ed. University of Utah Press.

Good, Irene L.
2001 Archaeological Textiles: A Review of Current Research. *Annual Review of Anthropology* 30:209–226.

Gould, Richard, and P. J. Watson
1982 A dialogue on the meaning and uses of analogy. *Journal of Anthropological Archaeology* 1:355–381.

Grifiths, D. R., et al.
1987 Experimental investigations of heat treatment of flint. In: *The Human Uses of Flint and Chert*. Sieveking, G. de G. and M. H. Newcomer, eds. Proceedings of the Fouth International Flint Symposium. Cambridge: Cambridge University Press.

Grissom, Carol A.
2000 Neolithic Statues from 'Ain Ghazal: Construction and Form. *American Journal of Archaeology* 104:25–45.

Hamilton, Elizabeth G.
1996 *Technology and Social Change in Belgic Gaul: Copper Working at the Titelberg, Luxembourg, 125 BC–AD 300*. MASCA Research Papers in Science and Archaeology, Vol. 13. Philadelphia: University of Pennsylvania Museum.

Hecht, Ann
1989 *The Art of the Loom. Weaving, Spinning and Dyeing across the World*. London: British Museum Publications.

Hegmon, Michelle
1992 Archaeological Research on Style. *Annual Review of Anthropology* 21:517–536.

Hegmon, Michelle
1998 Technology, Style, and Social Practices: Archaeological Approaches. In: *The Archaeology of Social Boundaries*. Stark, Miriam T., ed. Smithsonian Series in Archaeological Inquiry. Washington DC: Smithsonian Institution Press. pp. 264–279.

Helms, Mary W.
1993 *Craft and the Kingly Ideal. Art, Trade, and Power*. Austin: University of Texas Press.

Henderson, Julian
2000 *The Science and Archaeology of Materials. An investigation of inorganic materials*. London: Routledge.

Hester, Thomas R., Harry J. Shafer, and Kenneth L. Feder
1997 *Field Methods in Archaeology*. 7th ed. Mountain View, California: Mayfield Publishers.

Heyerdahl, Thor
1980 *The Tigris Expedition: In search of our beginnings*. London: Allen and Unwin.

Hietala, Harold J., ed.
1984 *Intrasite Spatial Analysis in Archaeology*. Cambridge: Cambridge University Press.

Hillman, Gordon C.
1984 Interpretation of Archaeological Plant Remains: The Application of Ethnographic Models from Turkey. In: *Plants and Ancient Man—Studies in Paleoethnobotany*. Van Zeist, W. and W. A. Casparie, eds. Rotterdam: A. A. Balkema. pp. 1–41.

Hodder, Ian
1982 *The Present Past*. London: Batsford.

Hodges, Henry
1970 *Technology in the Ancient World*. (reprinted in 1992 by Barnes and Noble). London: Allen Lane The Penguin Press.

Hodges, Henry
1989 [1976] *Artifacts: An introduction to early materials and technology*. Reprint of second edition. First edition published by John Baker, London, in 1964; 2nd edition published in 1976. London: Gerald Duckworth & Co. Ltd.

Horne, Lee
1982 Fuel for the Metal Worker. The Role of Charcoal and Charcoal Production in Ancient Metallurgy. *Expedition* 25(1):6–13.

Horne, Lee
1987 The Brasscasters of Dariapur, West Bengal. Artisans in a Changing World. *Expedition* 29(3):39–46.

Horne, Lee
1990 A Study of Traditional Lost-Wax Casting in India. *JOM (The journal of the Minerals, Metals & Materials Society)* 42(10):46–47.

Hosler, Dorothy
1994a Sound, color and meaning in the metallurgy of Ancient West Mexico. *World Archaeology* 27(1):100–115.

Hosler, Dorothy
1994b *The Sounds and Colors of Power: The Sacred Metallurgical Technology of Ancient West Mexico*. Cambridge, MA: Massachusetts Institute of Technology (MIT) Press.

Hruby, Zachary X., and Rowan Flad, eds.
2006 *Rethinking Specialization in Complex Societies: Archaeological Analysis of the Social Meaning of Production*. Archaeological Papers of the American Anthropological Association (AP3A), Volume 15. Berkeley, CA: University of California Press and the American Anthropological Association.

Hudson, Travis, and T.C. Blackburn
1982 *The Material Culture of the Chumash Interaction Sphere. Vol. I: Food Procurement and Transportation*. Ballena Press Anthropological Papers, No. 27. Los Altos and Santa Barbara, CA: Ballena Press and Santa Barbara Museum of Natural History.

Hudson, Travis, Janice Timbrook, and Melissa Rempe (editors & annotators)
1978 *Tomol: Chumash Watercraft as Described in the Ethnographic Notes of John P. Harrington*. Ballena Press Anthropological Papers, No. 9. Socorro, NM: Ballena Press and Santa Barbara Museum of Natural History.

Hunter, Dard
1978 [1947] *Papermaking. The History and Technique of an Ancient Craft*. Unabridged republication of the second edition published by Alfred A. Knopf in 1947. New York, NY: Dover Publications.

Inizan, Marie-Louise, Hélène Roche, and Jaques Tixier
1992 *Technology of Knapped Stone. Followed by a multilingual vocabulary (Arabic, English, French, German, Greek, Italian, Russian, Spanish)*. Lee, Alan. Préhistoire de la Pierre Taillée, Tome 3. Originally published in French in 1980 as *Préhistoire de la Pierre Taillée* by Tixier, Inizan, and Roche. Meudon: Cercle de Recherches et d'Etudes Préhistoriques (CREP) and Centre National de las Recherche Scientifique (CNRS).

Jakes, Kathryn A., ed.
2002 *Archaeological Chemistry: Materials, methods, and meaning*. Washington D.C.: American Chemical Society, distributed by Oxford University Press.

James, Peter, and Nick Thorpe
1994 *Ancient Inventions*. New York, NY: Ballantine Books.

Johnston, Robert H.
1977 The Development of the Potter's Wheel: An Analytical and Synthesizing Study. In: *Material Culture: Styles, Organization and Dynamics of Technology*. Lechtman, Heather and Robert S. Merrill, eds. Proceedings of the American Ethnological Society. St. Paul, MN: West Publishing Co. pp. 169–210.

Johnstone, Paul
1980 *The Sea-Craft of Prehistory*. Cambridge, MA: Harvard University Press.

Jones, Glynnis E. M.
1987 A Statistical Approach to the Archaeological Identification of Crop Processing. *Journal of Archaeological Science* 14:311–323.

Jørgensen, Lise Bender
1992 *Northern European Textiles until AD 1000*. Århus, Denmark: Aarhus University Press.

Juleff, Gillian
1998 *Early Iron and Steel in Sri Lanka. A Study of the Samanalawewa Area*. Materialien zur Allgemeinen und Vergleichenden Archäologie, Band (Volume) 54. Mainz am Rhein: Verlag Philipp von Zabern.

Kardulias, P. Nick, and Richard W. Yerkes
2003 Introduction: Lithic Analysis as Cross-Cultural Study. In: *Written in Stone. The Multiple Dimensions of Lithic Analysis.* Kardulias, P. Nick and Richard W. Yerkes, eds. Lanham, Maryland: Lexington Books. pp. 1–5.

Kenoyer, Jonathan Mark
1991 Ornament Styles of the Indus Tradition: Evidence from Recent Excavations at Harappa, Pakistan. *Paléorient* 17(2):79–98 (published in 1992 for 1991).

Kenoyer, Jonathan Mark
1992 Harappan Craft Specialization and the Question of Urban Segregation and Stratification. *Eastern Anthropologist. Indus Civilization Special Number* 45(1–2):39–54.

Kenoyer, Jonathan Mark
1994 Experimental studies of Indus Valley technology at Harappa. In: *South Asian Archaeology 1993.* Parpola, Asko and Petteri Koskikallio, eds. Annales Academiae Scientiarum Fennicae, Series B, Volume 271, Vol. 1. Helsinki: Suomalainen Tiedeakatemia. pp. 345–362.

Kenoyer, Jonathan Mark
1995 Shell Trade and Shell Working During the Neolithic and Early Chalcolithic at Mehrgarh, Pakistan. In: *Mehrgarh: Field Reports 1974–1985 from Neolithic Times to the Indus Civilization.* Jarrige, Catherine, Jean-François Jarrige, Richard H. Meadow, and Gonzague Quivron, eds. Karachi: Dept. of Culture & Tourism, Government of Sindh, Pakistan, in collaboration with the French Ministry of Foreign Affairs. pp. 566–581.

Kenoyer, Jonathan Mark
1998a *Ancient Cities of the Indus Valley Civilization.* Karachi: Oxford University Press and American Institute of Pakistan Studies.

Kenoyer, Jonathan Mark
1998b Wealth and Socio-Economic Hierarchies of the Indus Valley Civilization. In: *Order, Legitimacy and Wealth in Early States.* Richards, Janet and Mary Van Buren, eds. Cambridge: Cambridge University Press. pp. 88–109.

Kenoyer, Jonathan Mark
n.d. *Ancient Technology: From Stone Tools to Metallurgy.* Unpublished handbook for Ancient Technology and Invention Lab, Dept. of Anthropology, University of Wisconsin-Madison.

Kenoyer, Jonathan Mark, Massimo Vidale, and Kuldeep Kumar Bhan
1991 Contemporary stone bead-making in Khambhat, India: patterns of craft specialization and organization of production as reflected in the archaeological record. *World Archaeology* 23(1):44–63.

Kenoyer, Jonathan Mark, Massimo Vidale, and Kuldeep Kumar Bhan
1994 Carnelian Bead Production in Khambhat, India: An Ethnoarchaeological Study. In: *Living Traditions. Studies in the Ethnoarchaeology of South Asia.* Allchin, Bridget, ed. New Delhi: Oxford and IBH Publishing Co. Pvt. Ltd. pp. 281–306.

Killen, Geoffrey
1994 *Egyptian Woodworking and Furniture.* Shire Egyptology, Number 21. Princes Risborough, Buckinghamshire, UK: Shire Publications Ltd.

Kingery, W. David, ed.
1985 *Technology and Style.* Ceramics and Civilization, Volume II. Columbus, OH: The American Ceramic Society.

Kingery, W. David
1986 The Development of European Porcelain. In: *High-Technology Ceramics: Past, Present, and Future. The Nature of Innovation and Change in Ceramic Technology.* Kingery, W. David, ed. Ceramics and Civilization, Volume III. Columbus, OH: The American Ceramic Society. pp. 153–180.

Kingery, W. David
1993 Technological Systems and some Implications with Regard to Continuity and Change. In: *History from Things: Essays on Material Culture.* Lubar, Steven and W. David Kingery, eds. Washington DC: Smithsonian Institution Press. pp. 215–230.

Kingery, W. David, ed.
1996 *Learning from Things: Method and Theory of Material Culture Studies.* Washington D.C.: Smithsonian Institution Press.

Kingery, W. David
2001 The Design Process as a Critical Component of the Anthropology of Technology. In: *Anthropological Perspectives on Technology.* Schiffer, Michael B., ed. Albuquerque: University of New Mexico Press. pp. 123–138.

Kingery, W. David, and W. Patrick McCray, eds.
1998 *The Prehistory and History of Glass Technology.* Ceramics and Civilization Series, Volume VIII. Columbus, OH: The American Ceramic Society.

Kingery, W. David, Pamela B. Vandiver, and Martha Prickett.
1988 The Beginnings of Pyrotechnology, Part I: Production and Use of Lime and Gypsum Plaster in the Pre-Pottery Neolithic Near East. *Journal of Field Archaeology* 15:219–244.

Klein, Richard G., and Kathryn Cruz-Uribe
1984 *The Analysis of Animal Bones from Archaeological Sites.* Prehistoric archaeology and ecology. Chicago: University of Chicago Press.

Knecht, Heidi
1997 Projectile Points of Bone, Antler and Stone. Experimental explorations of manufacture and use. In: *Projectile Technology.* Knecht, Heidi, ed. New York: Plenum Press. pp. 191–212.

Kroll, Ellen M., and T. Douglas Price, eds.
1991 *The Interpretation of Archaeological Spatial Patterning.* New York: Plenum Press.

Lahiri, Nayanjot
1993 Some Ethnographic Aspects of the Ancient Copper-Bronze Tradition in India. *Journal of the Royal Asiatic Society, Series 3*, 3(2):219–231.

Lahiri, Nayanjot
1995 The Alloying Traditions of Protohistoric and Historic India. Some General Trends and Their Ethnographic Dimensions. *Puratattva* 25 (for 1994/95):64–73.

Latta, Martha A., Paul Thibaudeau, and Lisa Anselmi
1998 Expediency and Curation: The Use and Distribution of "Scrap" Trade Metal by Huron Native Peoples in Sixteenth-Century Southern Ontario. *Wisconsin Archaeologist* 79(1):175–184.

Law, Randall
2005 Regional Interaction in the Prehistoric Indus Valley: Initial Results of Rock and Mineral Sourcing Studies at Harappa. In: *South Asian Archaeology 2001. Volume I: Prehistory*. Jarrige Catherine and Vincent Lefèvre, eds. Paris: Éditions Recherche sur les Civilisations. pp. 179–190.

Law, Randall
2006 Moving Mountains: The Trade and Transport of Rocks and Minerals Within the Greater Indus Valley Region. In: *Space and Spatial Analysis in Archaeology*. Robertson, E.C., J.D. Seibert, D.C. Fernandez, and M.U. Zender, eds. Calgary, AL: Univ. of Calgary Press and Univ. of New Mexico Press. pp. 301–313.

Leader, J. M.
1991 The South Florida Metal Complex: A Preliminary Discussion of the Effects of the Introduction of an Elite Metal on a Contact Period Native American Society. In: *Metals in Society: Theory Beyond Analysis*. Ehrenreich, R.M., ed. MASCA Volume 8. Philadelphia, PA: University of Pennsylvania. pp. 19–24.

Lechtman, Heather
1976 A Metallurgical Site Survey in the Peruvian Andes. *Journal of Field Archaeology* 3(1):1–42.

Lechtman, Heather
1977 Style in Technology – Some Early Thoughts. In: *Material Culture: Styles, Organization and Dynamics of Technology*. Lechtman, Heather and Robert S. Merrill, eds. Proceedings of the American Ethnological Society. St. Paul, MN: West Publishing Co. pp. 3–20.

Lechtman, Heather
1980 The Central Andes: Metallurgy without Iron. In: *The Coming of the Age of Iron*. Wertime, Theodore A. and James D. Muhly, eds. New Haven, CT: Yale University Press. pp. 267–334.

Lechtman, Heather
1988 Traditions and Styles in Central Andean Metalworking. In: *The Beginning of the Use of Metals and Alloys*. Maddin, Robert, ed. Cambridge, MA, USA: Massechussetts Institute of Technology (MIT) Press. pp. 344–378.

Lechtman, Heather
1999 Afterword. In: *The Social Dynamics of Technology. Practice, Politics, and World Views*. Dobres, Marcia-Anne and Christopher R. Hoffman, eds. Washington DC: Smithsonian Institution Press. pp. 223–232.

Lechtman, Heather, and Robert S. Merrill
1977 *Material Culture: Styles, Organization and Dynamics of Technology*. Proceedings of the American Ethnological Society. St. Paul, MN: West Publishing Co.

Lechtman, Heather, and Arthur Steinberg
1979 The History of Technology: An Anthropological Point of View. In: *The History and Philosophy of Technology*. Bugliarello, G. and D. B. Doner, eds. Urbana: University of Illinois Press. pp. 135–160.

LeMoine, Genevieve
2001 Skeletal Technology in Context: An Optimistic Overview. In: *Crafting Bone: Skeletal Technologies through Space and Time*. Choyke, Alice M. and László Bartosiewicz, eds. BAR International Series 937. Oxford: Archaeopress. pp. 1–7.

Lemonnier, Pierre
1986 The study of material culture today: toward an anthropology of technical systems. *Journal of Anthropological Archaeology* 5:147–186.

Lemonnier, Pierre
1992 *Elements for an Anthropology of Technology*. Anthropological Papers, Museum of Anthropology, Univ. of Michigan, No. 88. Ann Arbor, Michigan: Museum of Anthropology, University of Michigan.

Lemonnier, Pierre
1993 Introduction. In: *Technological Choices: Transformation in material cultures since the Neolithic*. Lemonnier, Pierre, ed. London: Routledge. pp. 1–35.

Lesure, Richard
1999 On the Genesis of Value in Early Hierarchical Societies. In: *Material Symbols: Culture and Economy in Prehistory*. Robb, John E., ed. Occasional Paper No. 26, Visiting Scholar Program. Carbondale, IL: Center for Archaeological Investigations, Southern Illinois University. pp. 23–55.

Leute, Ulrich
1987 *Archaeometry: An introduction to physical methods in archaeology and the history of art*. Weinheim, FDR (Germany): VCH Publishers.

Lock, Gary, ed.
2000 *Beyond the Map: Archaeology and spatial technologies*. NATO Science series. Series A, Life Sciences. Vol. 321. Amsterdam & Washington DC: IOC Press.

Lowe, Thelma L.
1989a Refractories in High-Carbon Iron Processing: A Preliminary Study of the Deccani Wootz-Making Crucibles. In: *Cross-Craft and Cross-Cultural Interactions in Ceramics*. McGovern, Patrick E. and M. D. Notis, eds. Ceramics and Civilization Series, Volume IV. Westerville, OH: The American Ceramic Society, Inc. pp. 237–251.

Lowe, Thelma L.
1989b   Solidification and the Crucible Processing of Deccani Ancient Steel. In: *Principles of Solidification and Materials Processing*. Trivedi, R., J. A. Sekhar, and J. Mazumdar, eds. New Delhi: Oxford & IBH Publishing. pp. 729–740.

Lubar, Steven, and W. David Kingery, eds.
1993   *History from Things: Essays on Material Culture*. Washington DC: Smithsonian Institution Press.

Luedtke, Barbara E.
1992   *An Archaeologist's Guide to Chert and Flint*. Archaeological Research Tools 7. Los Angeles: Institute of Archaeology, UCLA.

Luedtke, Barbara E.
1999a   Gunflints in the Northeast. *Northeast Anthropology* 57:27–43.

Luedtke, Barbara E.
1999b   What makes a good gunflint? *Archaeology of Eastern North America* 27:71–79.

MacGregor, Arthur
1985   *Bone, Antler, Ivory and Horn: The Technology of Skeletal Materials since the Roman Period*. Croom Helm, London, and Barnes and Noble Books, Totowa, New Jersey.

Mackay, Ernest J. H.
1931   Ch. XXVIII. Ivory, Shell, Faience, and Other Objects of Technical Interest. In: *Mohenjo-daro and the Indus Civilization*. Marshall, John, ed, Vol. 2. London: Arthur Probsthain. pp. 562–588.

Mackay, Ernest J. H.
1938   *Further Excavations at Mohenjo-daro*. Delhi: Government of India Press.

MacKenzie, Maureen Anne
1991   *Androgynous Objects: String bags and gender in central New Guinea*. Philadelphia: Harwood Academic Publishers.

Martin, Susan R.
1999   *Wonderful Power: The Story of Ancient Copper Working in the Lake Superior Basin*. Detroit, Michigan: Wayne State University Press.

Matarasso, Pierre, and Valentine Roux
2000   Le système techno-économique des perles de corneline. Modélisation des systèmes complexes de production par l'analyse d'activités (The Techno-system of carnelian beads. Modelling of complex systems of production by the analysis of activities). In: *Cornaline de l'Inde. Des pratiques techniques de Cambay aux techno-systèmes de l'Indus*. Roux, Valentine, ed. Paris: Éditions de la Maison des sciences de l'homme. pp. 333–410.

Mathieu, James R.
2002   Introduction—Experimental Archaeology: Replicating Past Objects, Behaviors, and Processes. In: *Experimental Archaeology: Replicating past objects, behaviors,*

and processes. Mathieu, James R., ed. BAR International Series 1035. Oxford, England: Archaeopress. pp. 1–12.

May, P., and M. Tuckson
1982 *The Traditional Pottery of Papua New Guinea.* Sydney: Bay Books.

McCarthy, Blythe, and Pamela Vandiver
1991 Ancient High-Strength Ceramics: Fritted Faience Bracelet Manufacture at Harappa (Pakistan), ca. 2300-1800 B.C. In: *Materials Issues in Art & Archaeology II: Symposium held April 17–21, 1990, San Francisco, CA.* Vandiver, Pamela et al., ed. Materials Research Society Symposium Proceedings, V. 185. Pittsburgh: Materials Research Society. pp. 495–510.

McCray, W. Patrick
1998 Glassmaking in Renaissance Italy: The Innovation of Venetian Cristallo. *JOM (The journal of the Minerals, Metals & Materials Society)* 50(5):14–19.

McCreight, Tim
1982 *The Complete Metalsmith: An Illustrated Handbook.* Worcester, MA: Davis Publishing.

McCreight, Tim
1986 *Practical Casting: A Studio Reference.* Boylston, MA: Brynmorgen Press.

McGhee, Robert
1977 Ivory for the Sea Woman: The Symbolic Attributes of a Prehistoric Technology. *Canadian Journal of Archaeology* 1:141–149.

McGovern, Patrick E., and M. D. Notis, eds.
1989 *Cross-Craft and Cross-Cultural Interactions in Ceramics.* Ceramics and Civilization, Volume IV. Westerville, OH: The American Ceramic Society.

McGrail, Sean
1985 Towards a classification of water transport. *World Archaeology* 16(3):289–303.

McMillion, Bill
1991 *The Archaeology Handbook: A field manual and resource guide.* New York: Wiley.

Merrill, Robert S.
1968 The Study of Technology. *International Encyclopedia of the Social Sciences* 15:577.

Merrill, Robert S.
1977 Preface. In: *Material Culture: Styles, Organization and Dynamics of Technology.* Lechtman, Heather and Robert S. Merrill, eds. Proceedings of the American Ethnological Society. St. Paul, MN: West Publishing Co. pp. v–vii.

Méry, Sophie
1994 Excavation of an Indus potter's workshop at Nausharo (Baluchistan), Period II. In: *South Asian Archaeology 1993.* Parpola, Asko and Petteri Koskikallio, eds. Annales Academiae Scientiarum Fennicae, Series B, Volume 271, Vol. 2. Helsinki: Suomalainen Tiedeakatemia. pp. 471–482.

Méry, Sophie
1996 Ceramics and patterns of exchange across the Arabian Sea and the Persian Gulf in the Early Bronze Age. In: *The Prehistory of Asia and Oceania*. XIII International Congress of Prehistoric and Protohistoric Sciences, Forli, Italia, 8–14 September 1996. Volume 16. Afanas'ev, G., S. Cleuziou, John R. Lukacs, and Maurizio Tosi, eds. Forli, Italy: ABACO. pp. 167–179.

Méry, Sophie
2000 *Les céramiques d'Oman et l'Asie moyenne. Une archéologie des échanges à l'Âge du Bronze.* Collection de Recherches Archéologiques (CRA) Monographies, No. 23. Paris: Centre National de la Recherche Scientifique (CNRS) Éditions.

Miller, Daniel
1985 *Artefacts as Categories.* Cambridge: Cambridge University Press.

Miller, Heather M.-L.
1997 Pottery Firing Structures (Kilns) of the Indus Civilization During the Third Millennium B.C. In: *Prehistory and History of Ceramic Kilns*. Rice, Prudence and W. David Kingery, eds. Ceramics and Civilization, Volume VII. Columbus, OH: The American Ceramic Society, Inc. pp. 41–71.

Miller, Heather M.-L.
1999 *Pyrotechnology and Society in the Cities of the Indus Valley.* PhD thesis, University of Wisconsin–Madison. Available through UMI Digital Dissertations.

Miller, Heather M.-L.
2000 Reassessing the Urban Structure of Harappa: Evidence from Craft Production Distribution. In: *South Asian Archaeology 1997*. Taddei, Maurizio and Giuseppe De Marco, eds. Rome: Istituto Italiano per l'Africa e l'Oriente (IsIAO) & Istituto Universitario Orientale, Naples. pp. 77–100.

Miller, Heather M.-L.
2006 Associations and Ideologies in the Locations of Urban Craft Production at Harappa, Pakistan (Indus Civilization). In: *Rethinking Specialization in Complex Societies: Archaeological Analysis of the Social Meaning of Production.* Hruby, Zachary X. and Rowan Flad, eds. Archaeological Publications of the American Anthropological Association, Volume 15. Berkeley, CA: University of California Press and the American Anthropological Association.

Miller, Heather M.-L.
in press-a The Indus Talc-Faience Complex: Types of Materials, Clues to Production. In: *South Asian Archaeology 1999*. van Kooij, Karel R., Ellen M. Raven, and Gregory L. Possehl, eds. Leiden, Netherlands: International Institute of Asian Studies (IIAS).

Miller, Heather M.-L.
in press-b Issues in the Determination of Ancient Value Systems: The Role of Talc (Steatite) and Faience in the Indus Civilization. In: *Intercultural Relations between South and Southwest Asia. Studies in Commemoration of E.C.L. During Caspers*

(1934–1996). Olijdam, Eric and R. H. Spoor, eds. Oxford: BAR International Series.

Millon, René
1973 *Urbanization at Teotihuacan, Mexico. Vol I: the Teotihuacan map. Part I: text.* Austin: University of Texas Press.

Millon, René, Bruce Drewitt, and George L. Cowgill
1973 *Urbanization at Teotihuacan, Mexico. Vol I: the Teotihuacan map. Part 2: maps.* Austin: University of Texas Press.

Moorey, P. R. S.
1994 *Ancient Mesopotamian Materials and Industries.* Oxford: Clarendon Press.

Mukherjee, Meera
1978 *Metalcraftsmen of India.* Anthropological Survey of India, Memoir No. 44. Calcutta: Government of India.

Murphy, Peter, and Patricia E. J. Wiltshire, eds.
2003 *The Environmental Archaeology of Industry.* Symposia of the Association for Environmental Archaeology No. 20. Oxford: Oxbow Books.

Nicholson, Paul T.
1993 *Egyptian Faience and Glass.* Shire Egyptology, Number 18. Princes Risborough, Buckinghamshire, UK: Shire Publications Ltd.

Nicholson, Paul T.
1998 Materials and Technology. In: *Gifts of the Nile. Ancient Egyptian Faience.* Friedman, Florence Dunn, ed., with the assistance of Georgina Borromeo. London: Thames and Hudson. pp. 50–64.

Nicholson, Paul T., with Edgar Peltenburg
2000 Egyptian Faience. In: *Ancient Egyptian Materials and Technology.* Nicholson, Paul T. and Ian Shaw, eds. Cambridge: Cambridge University Press. pp. 177–194.

Nicholson, Paul T., and Julian Henderson
2000 Egyptian Glass. In: *Ancient Egyptian Materials and Technology.* Nicholson, Paul T. and Ian Shaw, eds. Cambridge: Cambridge University Press. pp. 195–224.

O'Connor, Terence Patrick
2000 *The Archaeology of Animal Bones.* College Station, Texas: Texas A&M University Press.

Ochsenschlager, E.
1992 Ethnographic Evidence for Wood, Boats, Bitumen and Reeds in Southern Iraq: Ethnoarchaeology at al-Hiba. In: *Trees and Timber in Mesopotamia.* Postgate, J. N. and Mark A. Powell, eds. Bulletin on Sumerian Agriculture, Volume 6. Cambridge: Sumerian Agriculture Group, Faculty of Oriental Studies, University of Cambridge. pp. 47–78.

Odell, Goerge H.
2004 *Lithic Analysis*. Manuals in Archaeological Method, Theory, and Technique. New York: Kluwer Academic/Plenum Publishers.

Ogden, Jack M.
1992 *Interpreting the Past: Ancient Jewellery*. London: BMP.

Orser, Charles E.
2004 *Historical Archaeology*. Second edition. Upper Saddle River, NJ: Pearson/Prentice Hall.

Orton, Clive, Paul Tyers, and Alan Vince
1993 *Pottery in Archaeology*. Cambridge Manuals in Archaeology. Cambridge: Cambridge University Press.

Parsons, Jeffrey R., and Mary H. Parsons
1990 *Maguey Utilization in Highland Central Mexico. An Archaeological Ethnography*. Anthropological Papers of the Museum of Anthropology, University of Michigan, No. 82. Ann Arbor: Museum of Anthropology, University of Michigan.

Pearsall, Deborah M.
1989 *Paleoethnobotany: A handbook of procedures*. San Diego: Academic Press.

Pelegrin, Jaques
2000 Technique et méthodes de taille pratiquées à Cambay (The knapping methods and techniques practiced at Cambay). In: *Cornaline de l'Inde. Des pratiques techniques de Cambay aux techno-systèmes de l'Indus*. Roux, Valentine, ed. Paris: Éditions de la Maison des sciences de l'homme. pp. 53–93.

Pfaffenberger, Brian
1992 Social Anthropology of Technology. *Annual Review of Anthropology* 21:491–516.

Piel-Desruisseux, Jean-Luc
1998 *Outils préhistoriques. Formes, fabrication, utilisation*. Third edition. Paris: Masson.

Pigott, Vincent C., ed.
1999a *The Archaeometallurgy of the Asian Old World*. MASCA Research Papers in Science and Archaeology 16; University of Pennsylvania Museum Monograph 89. Philadelphia: The University Museum, University of Pennsylvania.

Pigott, Vincent C.
1999b The Development of Metal Production on the Iranian Plateau: An Archaeometallurgical Perspective. In: *The Archaeometallurgy of the Asian Old World*. Pigott, Vincent C., ed. MASCA Research Papers in Science and Archaeology 16; University of Pennsylvania Museum Monograph 89. Philadelphia: The University Museum, University of Pennsylvania. pp. 73–106.

Piperno, Dolores
1988 *Phytolith Analysis: An archaeological and geological perspective*. San Diego: Academic Press.

Plog, Stephen, and Julie Solometo
1997 The Never-Changing and the Ever-Changing: the Evolution of Western Pueblo Ritual. *Cambridge Archaeological Journal* 7(2):161–182.

Possehl, Gregory L.
1998 Sociocultural Complexity Without the State: The Indus Civilization. In: *Archaic States*. Feinman, Gary M. and Joyce Marcus, eds. Sante Fe, NM: School of American Research. pp. 261–291.

Pracchia, Stefano
1987 Surface Analysis of Pottery Manufacture Areas at Moenjodaro. The Season 1984. In: *Interim Reports Volume 2. Reports on Field Work Carried out at Mohenjo-Daro Pakistan 1983–84*. Jansen, Michael and Günter Urban, eds. Aachen and Rome: German Research Project Mohenjo-Daro, Rheinish-Westfälische Technische Hochschule (RWTH) Aachen, and Istituto Italiano per il Medio ed Estremo Oriente (IsMEO). pp. 151–166.

Pracchia, Stefano, Maurizio Tosi, and Massimo Vidale
1985 On the Type, Distribution and Extent of Craft Industries at Moenjo-daro. In: *South Asian Archaeology 1983*. Schotsmans, Janine and Maurizio Taddei, eds. Istituto Universitario Orientale, Dipartimento di Studi Asiatici, Series Minor XXIII. Naples: Istituto Universitario Orientale. pp. 207–247.

Pracchia, Stefano, and Massimo Vidale
1990 Understanding Pottery Production through the Identification and Analysis of Kiln Remains. *Prospezioni Archeologiche Quaderni* 1:63–68.

Provenzano, Noëlle
2001 Worked Bone Assemblages from Northern Italian Terramares: A Technological Approach. In: *Crafting Bone: Skeletal Technologies through Space and Time*. Choyke, Alice M. and László Bartosiewicz, eds. BAR International Series 937. Oxford: Archaeopress. pp. 93–101.

Purdy, Barbara A.
1982 Pyrotechnology: Prehistoric Applications to Chert Material. In: *Early Pyrotechnology: The Evolution of the First Fire-Using Industries*. Wertime, Theodore A. and Steven F. Wertime, eds. Washington DC: Smithsonian Institution Press. pp. 31–43.

Pye, Elizabeth
2001 *Caring for the Past: Issues in conservation for archaeology and museums*. London: James and James (Science Publishers).

Rawson, J.
1989 Chinese Silver and Its Influence on Porcelain Development. In: *Cross-Craft and Cross-Cultural Interactions in Ceramics*. McGovern, Patrick E. and M. D. Notis, eds. Ceramics and Civilization Series, Volume IV. Westerville, OH: The American Ceramic Society, Inc. pp. 275–300.

Reddy, Seetha Narahari
1997 If the Threshing Floor Could Talk: Integration of Agriculture and Pastoralism during the Late Harappan in Gujarat, India. *Journal of Anthropological Archaeology* 16:162–187.

Reents-Budet, Dorie
1998 Elite Maya Pottery and Artisans as Social Indicators. In: *Craft and Social Identity.* Costin, Cathy L. and Rita P. Wright, eds. Archaeological Papers of the American Anthropological Association (AP3A), Number 8. Arlington, VA: American Anthropological Association. pp. 71–89.

Reeves, Ruth
1962 *Ciré Perdue Casting in India.* New Delhi: Crafts Museum.

Rehren, Thilo, Edgar B. Pusch, and Anja Herold
2001 Qantir-Piramesses and the organisation of the Egyptian glass industry. In: *The Social Context of Technological Change: Egypt and the Near East, 1650–1550 BC.* Shortland, Andrew, ed. Proceedings of a conference held at St. Edmund Hall, Oxford, 12–14 September 2000. Oxford: Oxbow. pp. 223–238.

Reitz, Elizabeth Jean, and Elizabeth S. Wing
1999 *Zooarchaeology.* Cambridge Manuals in Archaeology. Cambridge: Cambridge University Press.

Renfrew, Colin, and Paul Bahn
2000 *Archaeology: Theories, Methods and Practice.* Third edition. New York: Thames and Hudson Inc.

Rhodes, Daniel
1968 *Kilns. Design, Construction, and Operation.* Philadelphia, PA: Chilton Book Company.

Rice, Prudence M.
1987 *Pottery Analysis: A Sourcebook.* Chicago: University of Chicago Press.

Rice, Prudence M.
1996a Recent Ceramic Analysis: 1. Function, Style, and Origins. *Journal of Archaeological Research* 4(2):133–163.

Rice, Prudence M.
1996b Recent Ceramic Analysis: 2. Composition, Production, and Theory. *Journal of Archaeological Research* 4(3):165–202.

Rice, Prudence M.
2000 *Integrating Archaeometry.* Occasional Paper No. 29, Visiting Scholar Program. Carbondale, IL: Center for Archaeological Investigations, Southern Illinois University.

Rissman, Paul
1988 Public displays and private values: a guide to buried wealth in Harappan archaeology. *World Archaeology* 20(2):209–228.

Rolland, Nicolas, and Harold L. Dibble
1990 A new synthesis of Middle Paleolithic assemblage variability. *American Antiquity* 55(3):480–499.

Roskams, Steve
2001 *Excavation*. Cambridge Manuals in Archaeology. Cambridge: Cambridge University Press.

Rothenberg, Beno, and Antonio Blanco-Freijeiro
1981 *Studies in Ancient Mining and Metallurgy in South-West Spain. Explorations and Excavations in the Province of Huelva*. Metal in History, Volume 1. London: IAMS (Institute for Archaeo-Metallurgy Studies).

Roux, Valentine, ed.
2000 *Cornaline de l'Inde. Des pratiques techniques de Cambay aux techno-systèmes de l'Indus*. Paris: Éditions de la Maison des sciences de l'homme.

Roux, Valentine
2003 A Dynamic Systems Framework for Studying Technological Change: Application to the Emergence of the Potter's Wheel in the Southern Levant. *Journal of Archaeological Method and Theory* 10(1):1–30.

Roux, Valentine, Blandine Bril, and Gilles Dietrich
1995 Skills and learning difficulties involved in stone knapping: the case of stone-bead knapping in Khambhat, India. *World Archaeology* 27(1):63–87.

Roux, Valentine, and Pierre Matarasso
1999 Crafts and the Evolution of Complex Societies: New Methodologies for Modeling the Organization of Production, a Harappan Example. In: *The Social Dynamics of Technology. Practice, Politics, and World View*. Dobres, Marcia-Anne and Christopher R. Hoffman, eds. Washington DC: Smithsonian Institution Press. pp. 46–70.

Russell, Nerissa
2001a Neolithic Relations of Production: Insights from the Bone Tool Industry. In: *Crafting Bone: Skeletal Technologies through Space and Time*. Choyke, Alice M. and László Bartosiewicz, eds. BAR International Series 937. Oxford: Archaeopress. pp. 271–280.

Russell, Nerissa
2001b The Social Life of Bone: A Preliminary Assessment of Bone Tool Manufacture and Discard at Çatalhöyük. In: *Crafting Bone: Skeletal Technologies through Space and Time*. Choyke, Alice M. and László Bartosiewicz, eds. BAR International Series 937. Oxford: Archaeopress. pp. 241–249.

Rye, Owen S.
1976 Appendix 3. Khar: Sintered Plant Ash in Pakistan. In: *Traditional Pottery Techniques of Pakistan. Field and Laboratory Studies*. Rye, Owen S. and Clifford Evans, eds. Smithsonian Contributions to Anthropology, Number 21. Washington DC: Smithsonian Institution Press. pp. 180–185.

Rye, Owen S.
1981 *Pottery Technology: Principles and Reconstruction.* Manuals on Archeology No. 4. Washington DC: Taraxacum Press.

Sackett, James R.
1986 Isochrestism and Style: A Clarification. *Journal of Anthropological Archaeology* 5:266–277.

Sackett, James R.
1990 Style and Ethnicity in Archaeology: The Case for Isochrestism. In: *The Uses of Style in Archaeology.* Conkey, Margaret W. and Christine A. Hasdorf, eds. Cambridge: Cambridge University Press. pp. 32–43.

Sands, Rob
1997 *Prehistoric Woodworking. The Analysis and Interpretation of Bronze and Iron Age Toolmarks.* Wood in Archaeology, Volume 1. London: Institute of Archaeology, University College London.

Scheel, Bernd
1989 *Egyptian Metalworking and Tools.* Shire Egyptology. Aylesbury, Bucks, UK: Shire Publications Ltd.

Schibler, Jörg
2001 Experimental Production of Neolithic Bone and Antler Tools. In: *Crafting Bone: Skeletal Technologies through Space and Time.* Choyke, Alice M. and László Bartosiewicz, eds. BAR International Series 937. Oxford: Archaeopress. pp. 49–60.

Schiffer, Michael B., ed.
2001 *Anthropological Perspectives on Technology.* Albuquerque: University of New Mexico Press.

Schiffer, Michael B., and A. R. Miller
1999 *The Material Life of Human Beings: Artifacts, Behavior and Communication.* London: Routledge.

Schiffer, Michael B., and James M. Skibo
1987 Theory and Experiment in the Study of Technological Change. *Current Anthropology* 28:595–622.

Schiffer, Michael B., James M. Skibo, Tamara C. Boelke, Mark A. Neupert, and Meredith Aronson
1994 New Perspectives on Experimental Archaeology: Surface Treatments and Thermal Response of the Clay Cooking Pot. *American Antiquity* 59:197–217.

Schmandt-Besserat, Denise
1980 Ocher in Prehistory: 300,000 Years of the Use of Iron Ores as Pigments. In: *The Coming of the Age of Iron.* Wertime, Theodore A. and James D. Muhly, eds. New Haven, CT: Yale University Press. pp. 127–150.

Schmidt, Armin
2002 *Geophysical Data in Archaeology: A guide to good practice.* 2nd edition. Archaeology Data Service. Oxford: Oxbow Books, on behalf of the Arts and Humanities Data Service.

Schneider, Gerwulf
1987 Chemical Analysis of Stoneware Bangles and Related Material from Mohenjo-Daro. In: *Interim Reports Volume 2. Reports on Field Work Carried out at Mohenjo-Daro Pakistan 1983–84.* Jansen, Michael and Günter Urban, eds. Aachen and Rome: German Research Project Mohenjo-Daro, Rheinish-Westfälische Technische Hochschule (RWTH) Aachen, and Istituto Italiano per il Medio ed Estremo Oriente (IsMEO). pp. 73–77.

Schwartz, Mark, and David Hollander
2001 Annealing, Distilling, Reheating and Recycling: Bitumen Processing in the Ancient Near East. *Paléorient* 26(2):83–91.

Schwartz, Mark, and David Hollander
2006 Boats, Bitumen and Bartering: The Use of a Utilitarian Good to Track Movement and Transport in Ancient Exchange Systems. In: *Space and Spatial Analysis in Archaeology.* Robertson, Elizabeth C., Jeffrey D. Seibert, Deepika C. Fernandez, and Mark U. Zender, eds. 34th Annual Conference of the University of Calgary Archaeological Association. Calgary, Alberta, Canada: University of Calgary Press and University of New Mexico Press. pp. 323–330.

Sciuti, Sebastiano, ed.
1996 *Notes on Archaeometry: Non-destructive testing techniques in archaeometry.* Rome: Bagatto.

Scollar, Irwin, A. Tabbagh, A. Hesse, and I. Herzog
1990 *Archaeological Prospecting and Remote Sensing.* Cambridge: Cambridge University Press.

Scott, David A.
1991 *Metallography and Microstructure of Ancient and Historic Metals.* The Getty Conservation Institute, J. Paul Getty Museum and Archetype Books.

Scott, David A.
2002 *Copper and Bronze in Art. Corrosion, Colorants, Conservation.* Los Angeles, CA: Getty Publications.

Sease, Catherine
1994 *A Conservation Manual for the Field Archaeologist.* 3rd edition. Archaeological Research Tools, Volume 4. Los Angeles: Institute of Archaeology, Unviersity of California Los Angeles.

Seiler-Baldinger, Annemarie
1994 *Textiles. A Classification of Techniques.* Washington, D.C.: Smithsonian Institution Press.

Seymour, John
1984   *The Forgotten Crafts*. American (English edition title "The Forgotten Arts".) New York: Random House.

Shackley, M. S., ed.
1997   *Archaeological Obsidian Studies: Method and Theory*. New York: Plenum Press.

Shah, Haku
1985   *Votive Terracottas of Gujarat*. Living Traditions of India. New York: Mapin International.

Sharer, Robert J., and Wendy Ashmore
2003   *Archaeology. Discovering Our Past*. Third edition. New York: McGraw Hill.

Shennan, Stephen
1997   *Quantifying Archaeology*. Edinburgh: Edinburgh University Press.

Shepard, Anna O.
1976   *Ceramics for the Archaeologist*. Washington D.C.: Carnegie Institution of Washington.

Shimada, Izumi, and Jo Ann Griffin
1994   Precious metal objects of the Middle Sicán. *Scientific American* 270(4):82–89.

Shimada, Izumi, and John F. Merkel
1991   Copper-alloy metallurgy in ancient Peru. *Scientific American* 265(1):62–75.

Shortland, Andrew J.
2000   *Vitreous Materials at Amarna: The production of glass and faience in 18th Dynasty Egypt*. BAR International Series 827. Oxford: Archaeopress.

Singer, Charles, E. J. Holmyard, and A. R. Hall; Trevor I. Williams; Richard Raper; eds.
1954–1978   *A History of Technology*. 7 volumes. Volumes 1–5 edited by Singer, Holmyard, and Hall; Volumes 6–7 edited by Williams; Volume 8 (Consolidated Indexes) compiled by Raper. Oxford: Clarendon Press.

Sinopoli, Carla M.
1991   *Approaches to Archaeological Ceramics*. New York: Plenum Press.

Sinopoli, Carla M.
1998   Identity and Social Action among South Indian Craft Producers of the Vijayanagara Period. In: *Craft and Social Identity*. Costin, Cathy L. and Rita P. Wright, eds. Archaeological Papers of the American Anthropological Association (AP3A), Number 8. Arlington, VA: American Anthropological Association. pp. 161–172.

Sinopoli, Carla M.
2003   *The Political Economy of Craft Production. Crafting Empire in South India, c. 1350–1650*. Cambridge, UK: Cambridge University Press.

Skibo, James M. and Michael B. Schiffer
2001 Understanding Artifact Variability and Change: A Behavioral Framework. In: *Anthropological Perspectives on Technology*. Schiffer, Michael B., ed. Albuquerque: University of New Mexico Press. pp. 139–149.

Sloane, Eric
1962 *Diary of an Early American Boy. Noah Blake 1805*. New York: Funk and Wagnalls.

Soffer, Olga, J. M. Adovasio, and D. C. Hyland
2000 The "Venus" Figurines. Textiles, Basketry, Gender, and Status in the Upper Paleolithic. *Current Anthropology* 41(4):511–537.

Solometo, Julie
1999 *The Context and Process of Historic Era Pueblo Mural Painting*. Masters thesis, Dept. of Anthropology, University of Michigan.

Solometo, Julie
2000 *Context and Meaning in Pueblo Mural Painting*. Manuscript of a paper presented at the Chacmool Conference, Anthropology Dept., University of Calgary.

Spriggs, Matthew, ed.
1984 *Marxist Perspectives in Archaeology*. New Directions in Archaeology Series. Cambridge: Cambridge University Press.

Stark, Miriam T.
1998 Technological Choices and Social Boundaries in Material Culture Patterning: An Introduction. In: *The Archaeology of Social Boundaries*. Stark, Miriam T., ed. Smithsonian Series in Archaeological Inquiry. Washington DC: Smithsonian Institution Press. pp. 1–11.

Stech, Tamara
1999 Aspects of Early Metallurgy in Mesopotamia and Anatolia. In: *The Archaeometallurgy of the Asian Old World*. Pigott, Vincent C., ed. MASCA Research Papers in Science and Archaeology 16; University of Pennsylvania Museum Monograph 89. Philadelphia: The University Museum, University of Pennsylvania. pp. 59–71.

Stein, Gil J., and M. James Blackman
1993 The Organizational Context of Specialized Craft Production in Early Mesopotamian States. *Research in Economic Anthropology* 14:29–59.

Steinberg, Arthur
1977 Technology and Culture: Technological Styles in the Bronzes of Shang China, Phrygia and Urnfield Central Europe. In: *Material Culture: Styles, Organization and Dynamics of Technology*. Lechtman, Heather and Robert S. Merrill, eds. Proceedings of the American Ethnological Society. St. Paul, MN: West Publishers. pp. 53–86.

Stewart, Hilary
1984 *Cedar: Tree of Life to the Northwest Coast Indians*. Vancouver: Douglas and McIntyre.

Sullivan, Lynne P. and S. Terry Childs
2003 *Curating Archaeological Collections: From the field to the repository.* Lanham, MD: Altamira Press.

Teague, Lynn S.
1998 *Textiles in Southwestern Prehistory.* Albuquerque: University of New Mexico Press.

Thesiger, Wilfred
1964 *The Marsh Arabs.* London: Longmans, Green and Co., Ltd.

Theunissen, Robert, Peter Grave, and Grahame Bailey
2000 Doubts on diffusion: challenging the assumed Indian origin of Iron Age agate and carnelian beads in Southeast Asia. *World Archaeology* 32(1):84–105.

Thomas, Ken D.
1986 Section 4.7 The Bangles. In: *Lewan and the Bannu Basin. Excavation and survey of sites and environments in North West Pakistan.* Allchin, F. Raymond, Bridget Allchin, Farzand A. Durrani, and M. Farid Khan, eds. BAR International Series 310. Oxford: B. A. R. pp. 145–156.

Thompson, E. A.
1952 *A Roman Reformer and Inventor, being a new text of the treatise De rebus bellicis.* With a translation and introduction by E.A. Thompson and a Latin index by Barbara Flower. Oxford: Clarendon Press.

Tite, Michael S.
1972 *Methods of Physical Examination in Archaeology.* Studies in Archaeological Science. London: London Seminar Press.

Tite, Michael S., Ian C. Freestone, N. D. Meeks, and Paul T. Craddock
1985 The examination of refractory ceramics from metal-production and metalworking sites. In: *The Archaeologist and the Laboratory.* Phillips, Patricia, ed. Council for British Archaeology Research Report 58. London: Council for British Archaeology. pp. 50–55.

Torrence, Robin, and Sander E. van der Leeuw
1989 Introduction: what's new about innovation? In: *What's New? A closer look at the process of innovation.* van der Leeuw, Sander E. and Robin Torrence, eds. One World Archaeology, Volume 14. London: Unwin Hyman. pp. 1–15.

Tosi, Maurizio, and Massimo Vidale
1990 The Slippery Streets of Moenjodaro. *Prospezioni Archeologiche Quaderni* 1:7–12.

Tylecote, Ronald F.
1974 Can Copper Be Smelted in a Crucible? *Journal of the Historical Metallurgical Society* 8(1):54.

Tylecote, Ronald F.
1980 Furnaces, Crucibles, and Slags. In: *The Coming of the Age of Iron.* Wertime, Theodore A. and James D. Muhly, eds. New Haven, CT: Yale University Press. pp. 183–223.

Tylecote, Ronald F.
1987   *The Early History of Metallurgy in Europe*. Longman Archaeology Series. London: Longman.

Underhill, Anne P.
2002   *Craft Production and Social Change in Northern China*. Fundamental Issues in Archaeology. New York: Kluwer Academic/Plenum Publishers.

Untracht, O.
1975   *Metal Techniques for Craftsmen*. New York: Doubleday.

van der Leeuw, Sander E.
1977   Toward a study of the economics of pottery making. In: *Ex Horreo*. Van Beek, B.L., R.W. Brandt, and W. Goeman-van Waateringe, eds. Amsterdam: University of Amsterdam. pp. 68–76.

van der Leeuw, Sander E.
1993   Giving the Potter a Choice: Conceptual aspects of pottery techniques. In: *Technological Choices: Transformation in material cultures since the Neolithic*. Lemonnier, Pierre, ed. London: Routledge. pp. 238–288.

van der Leeuw, Sander E., and Robin Torrence, eds.
1989   *What's New? A closer look at the process of innovation*. One World Archaeology, Volume 14. London: Unwin Hyman.

Vandiver, Pamela B., and C. G. Koehler
1985   Structure, Processing, Properties, and Style of Corinthian Transport Amphoras. In: *Technology and Style*. Kingery, W. David, ed. Ceramics and Civilization, Volume II. Columbus, OH: The American Ceramic Society. pp. 173–215.

Vandiver, Pamela B., Olga Soffer, Bohuslav Klima, and Jiri Svoboda
1989   The Origins of Ceramic Technology at Dolni Vestonice, Czechoslovakia. *Science* 246:1002–1008.

Vanzetti, Alessandro, and Massimo Vidale
1994   Formation processes of beads: defining different levels of craft skill among the early beadmakers of Mehrgarh. In: *South Asian Archaeology 1993*. Parpola, Asko and Petteri Koskikallio, eds. Annales Academiae Scientiarum Fennicae, Series B, Volume 271, Vol. 2. Helsinki: Suomalainen Tiedeakatemia. pp. 763–776.

Verdet-Fierz, Bernard, and Regula Verdet-Fierz
1993   *Willow Basketry*. Originally published in 1993 as *Anleitung Zum Flechten nit Weiden* by Paul Haupt. Loveland, CO: Interweave Press.

Vickers, Michael
1989   The Cultural Context of Ancient Greek Ceramics: An Essay in Skeuomorphism. In: *Cross-Craft and Cross-Cultural Interactions in Ceramics*. McGovern, Patrick E.

and M. D. Notis, eds. Ceramics and Civilization Series, Volume IV. Westerville, OH: The American Ceramic Society, Inc. pp. 45–63.

Vidale, Massimo
1989 Specialized Producers and Urban Elites: on the Role of Craft Industries in Mature Harappan Urban Contexts. In: *Old Problems and New Perspectives in the Archaeology of South Asia*. Kenoyer, Jonathan Mark, ed. Wisconsin Archaeological Reports, Volume 2. Madison, WI: Department of Anthropology, University of Wisconsin–Madison. pp. 171–181.

Vidale, Massimo
1990 Stoneware Industry of the Indus Civilization: An Evolutionary Dead-End in the History of Ceramic Technology. In: *The Changing Roles of Ceramics in Society: 26,000 B.P. to the Present*. Kingery, W. David, ed. Ceramics and Civilization, Volume V. Westerville, OH: The American Ceramic Society, Inc. pp. 231–254.

Vidale, Massimo
1992 *Produzione Artigianale Protostorica: Etnoarcheologia e Archeologia*. Padua: Universita degli Studi di Padova.

Vidale, Massimo
1995 Early Beadmakers of the Indus Tradition. The Manufacturing Sequence of Talc Beads at Mehrgarh in the 5th Millennium B.C. *East and West* 45(1–4): 45–80.

Vidale, Massimo
2000 *The Archaeology of Indus Crafts. Indus craftspeople and why we study them*. IsIAO Reports and Memoirs, Series minor, IV. Rome.

Vidale, Massimo, Jonathan Mark Kenoyer, and Kuldeep K. Bhan
1992 A Discussion of the Concept of "Chaîne Opératoire" in the Study of Stratified Societies: Evidence from Ethnoarchaeology and Archaeology. In: *Ethnoarcheologie: Justification, Problémes, Limites*. Gallay, Alan, ed. XII Rencontres Internationales d'Archéologie et d'Histoire d'Antibes. Juan-Le-Pins, France: Center de Recherches Archéologiques. pp. 181–194.

Vidale, Massimo, and Heather M.-L. Miller
2000 On the development of Indus technical virtuosity and its relation to social structure. In: *South Asian Archaeology 1997*. Taddei, Maurizio and Giuseppe De Marco, eds. Rome: Istituto Italiano per l'Africa e l'Oriente (IsIAO) & Istituto Universitario Orientale, Naples. pp. 115–132.

Vidale, Massimo, and G. M. Shar
1991 Zahr Muhra: Soapstone-cutting in Contemporary Baluchistan and Sind. *Annali (Estratto da Annali dell'Istituto Universitaro Orientale, Napoli)* 50(1):61–78.

Vosmer, Tom
2000 Ships in the ancient Arabian Sea: the development of a hypothetical reed-boat model. *Proceedings of the Seminar for Arabian Studies* 30:235–242.

Wake, Thomas A.
1999 Exploitation of Tradition: Bone Tool Production and Use at Colony Ross, California. In: *The Social Dynamics of Technology. Practice, Politics, and World Views*. Dobres, Marcia-Anne and Christopher R. Hoffman, eds. Washington DC: Smithsonian Institution Press. pp. 186–208.

Walker, William H.
2001 Ritual Technology in an Extranatural World. In: *Anthropological Perspectives on Technology*. Schiffer, Michael B., ed. Albuquerque: University of New Mexico Press. pp. 87–106.

Walton, Penelope, and John-Peter Wild, eds.
1990 *Textiles in Northern Archaeology. NESAT III: Textiles Symposium in York, 6–9 May 1987*. North European Symposium for Archaeological Textiles (NESAT), Volume 3. London: Archetype Publications.

Wattenmaker, Patricia
1998 Craft Production and Social Identity in Northwest Mesopotamia. In: *Craft and Social Identity*. Costin, Cathy L. and Rita P. Wright, eds. Archaeological Papers of the American Anthropological Association (AP3A), Number 8. Arlington, VA: American Anthropological Association. pp. 47–56.

Weiner, Annette B., and Jane Schneider, eds.
1989 *Cloth and the Human Experience*. Washington DC: Smithsonian Institution Press.

Weitzman, David
1980 *Traces of the Past: A field guide to industrial archaeology*. New York: Scribner.

Wendrich, Willeke
1999 *The World According to Basketry. An Ethno-archaeological Interpretation of Basketry Production in Egypt*. Leiden, Netherlands: Research School of Asian, African and Amerindian Studies (CNWS), Universiteit Leiden.

Weymouth, John W., and Robert Huggins
1985 Geophysical Surveying of Archaeological Sites. In: *Archaeological Geology*. Rapp, George and John A. Gifford, eds. New Haven, CN: Yale University Press. pp. 191–235.

Whallon, Robert, and James A. Brown, eds.
1982 *Essays on Archaeological Typology*. Kampsville Seminars in Archeology, Volume 1. Evanston, Illinois: Center for American Archaeology Press, Northwestern University.

Wheatley, David, and Mark Gillings
2002 *Spatial Technology and Archaeology: the archaeological applications of GIS*. London: Taylor & Francis.

White, Kenneth D.
1984 *Greek and Roman Technology*. Ithaca, New York: Cornell University Press (also London: Thames and Hudson Ltd).

Whittaker, John C.
1994 *Flintknapping: Making and Understanding Stone Tools.* Austin: University of Texas Press.

Wiessner, Polly
1990 Is There a Unity to Style? In: *The Uses of Style in Archaeology.* Conkey, Margaret W. and Christine A. Hasdorf, eds. Cambridge: Cambridge University Press. pp. 105–112.

Wilk, Richard R.
2001 Toward an Archaeology of Needs. In: *Anthropological Perspectives on Technology.* Schiffer, Michael B., ed. Albuquerque: University of New Mexico Press. pp. 107–122.

Wobst, H. Martin
1977 Stylistic behaviour and information exchange. In: *For the Director: Research Essays in Honour of James B. Griffin.* Cleland, C.E., ed. Anthropological Papers of the Museum of Anthropology, Volume 61. Ann Arbor, MI: University of Michigan. pp. 317–342.

Wright, Rita P.
1989a New Perspectives on Third Millennium B.C. Painted Grey Wares. In: *South Asian Archaeology 1985.* Frifelt, Karen and Per Sørensen, eds. Scandinavian Institute of Asian Studies Occasional Paper 4. London and Riverdale, MD: Curzon Press and The Riverdale Company. pp. 137–149.

Wright, Rita P.
1989b New Tracks on an Ancient Frontier. In: *Archaeological Thought in America.* Lamberg-Karlovsky, C. C., ed. Cambridge: Cambridge University Press. pp. 268–279.

Wright, Rita P.
1991 Patterns of Technology and the Organization of Production at Harappa. In: *Harappa Excavations 1986–1990: A Multidisciplinary Approach to Third Millennium Urbanism.* Meadow, Richard H., ed. Monographs in World Archaeology Number 3. Madison, WI: Prehistory Press. pp. 71–88.

Wright, Rita P.
1993 Technological Styles: Transforming a Natural Material into a Cultural Object. In: *History from Things: Essays on Material Culture.* Lubar, Steven and W. David Kingery, eds. Washington DC: Smithsonian Institution Press. pp. 242–269.

Wright, Rita P.
1996a Introduction: Gendered Ways of Knowing in Archaeology. In: *Gender and Archaeology.* Wright, Rita P., ed. Philadelphia, PA: University of Pennsylvania Press. pp. 1–19.

Wright, Rita P.
1996b Technology, Gender and Class: Worlds of Difference in Ur III Mesopotamia. In: *Gender and Archaeology.* Wright, Rita P., ed. Philadelphia, PA: University of Pennsylvania Press. pp. 79–110.

Wright, Rita P.
1998 Crafting social identity in Ur III Southern Mesopotamia. In: *Craft and Social Identity*. Costin, Cathy L. and Rita P. Wright, eds. Archaeological Papers of the American Anthropological Association (AP3A), Number 8. Arlington, VA: American Anthropological Association. pp. 57–70.

Wylie, Alison
1982 An anology by any other name is just as analogical. *Journal of Anthropological Archaeology* 1:382–401.

Wylie, Alison
1985 The reaction against analogy. *Advances in Archaeological Method and Theory* 8:63–111.

Yellen, John E.
1977 *Archaeological Approaches to the Present: Models for reconstructing the past*. New York, NY: Academic Press.

Zickgraf, Benno
1999 *Geomagnetische und geoelecktrische Prospektion in der Archäologie: Systematick, Geschichte, Anwendung*. Rahden: M. Leidorf.